Cybersecurity Through Simulation

5 Wargaming Strategies for Protecting Business Infrastructure

Dr. Danny Bruna

ISBN Paperback: 978-1-967632-11-4

ISBN Hardcover: 978-1-967632-12-1

ISBN E-book: 978-1-967632-13-8

Dedication

I would like to dedicate this project to my ancestors, deities, and Jesus Christ, whose guidance and wisdom have been a constant source of inspiration. I extend my deepest gratitude to my family, whose unwavering support and encouragement have been instrumental in my journey. My mother's prayers, guidance, and love have been a beacon of strength. My father's stability and role-modeling have been exemplary for over forty years. It is important to note that my upbringing, culture, and family values have instilled in me the tools, lessons, and knowledge necessary to persevere.

Furthermore, I dedicate this work to the United States Army, whose rigorous discipline and unwavering standards shaped my resilience, work ethic, and dedication to excellence. Serving as an indirect fire mortar-man (11 Charlie) was both a challenge and an honor. It was filled with hard work, mental toughness, and discipline, which strengthened my resolve and commitment to the principles of freedom.

I would also like to extend my deepest gratitude to the United States Veterans Affairs, organizations supporting veterans, and the opportunities provided by the United States government. These institutions have supported my professional career, academic journey, and pursuit of a terminal degree.

Above all, I am proud to be a natural-born citizen of the United States. A nation that fosters innovation, education, and opportunity. It is within this land of freedom and endless possibilities that I have been able to pursue my dreams.

.

"Cybersecurity is a continuous cycle of protection, detection, response, and recovery."

– Chris Painter.

Acknowledgement

I would like to express my sincere gratitude to the individuals and institutions that have contributed to my academic journey. Firstly, I appreciate the support, motivation, and camaraderie from my fellow learners. Most importantly, I extend my gratitude to the faculty of the School of Business, Technology, and Healthcare Administration at Capella University in Minneapolis, Minnesota. Their guidance, mentorship, and support in the Doctor of Information Technology (DIT) program helped me achieve one of my dreams.

I also acknowledge the educators who laid the foundation for my academic success from the beginning of my studies, including the dedicated teachers from Walker Elementary School, William Dandy Middle School, Sunrise Middle School, Thurgood Marshall Elementary School, and Fort Lauderdale High School. These institutions were my community, foundation, and haven for acquiring knowledgeable, and lifelong skills. Notably, my computer science teacher at Fort Lauderdale High School inspired my interest in programming, mathematics, and computers. While my robotics teacher, math teacher, and English honors teachers provided invaluable guidance, insight, and mentorship. I can still remember one of my teachers lecturing about the knowledge of the Ancient Egyptians and their advanced mathematics at William Dandy Middle School. Moreover, it was the art teacher in elementary school and the art class that helped me discover my natural talent.

Furthermore, I appreciate the resources, safety, and support provided by the Broward County Main Library in Fort Lauderdale, Florida. The library served as a vital educational haven during my formative years, introducing me to the fundamentals of computer science, history, and self-improvement. I am proud to have persevered and achieved my goal of becoming a Doctor of Information Technology, specializing in cybersecurity. I am grateful for the contributions of each individual, as well as the ideas and institutions that have supported me along this life journey.

Table of Contents

List of Figures

List of Tables

Foreword

I met Dr. Daniel Bruna through our doctoral work at Capella University, where we both completed our doctorates in Information Technology. Like many of his colleagues, I had the opportunity to support Daniel throughout his dissertation journey. I helped him review his research and offered ideas on how to connect his academic work with a general audience, a task that's not always easy when dealing with topics such as cybersecurity and wargaming.

Daniel's book is the result of years of research, collaboration, and deep reflection on the evolving role of wargaming in cybersecurity. What makes this book especially valuable is that it does not just stay in the academic or theoretical space. Daniel not only breaks down the concepts of cyber wargaming but also makes them applicable and practical. He outlines real-world tools and strategies and shows how they can be applied across various domains, including education, business, space, and smart vehicle technologies. His ability to take complex topics and make them approachable is one of the things I admire most about this work.

Throughout the book, Daniel guides the reader through the application of wargaming in anticipating, preparing for, and managing cyber threats. He does this with clear language and thoughtful examples that make it easier to see how these strategies can be applied across different sectors. Whether you are looking to secure cyber-physical systems, understand influence strategies, or explore the expanding attack surfaces in aviation or AI, this book offers valuable insights.

This is a book I'm excited to bring into the classroom. It is relevant, timely, and offers practical insight for students and professionals alike. Daniel's passion for the subject is evident, and his work makes meaningful contributions to conversations around cybersecurity readiness. I believe it will resonate with anyone looking for actionable strategies in a rapidly changing digital landscape.

Serena Pontenila, PhD

Assistant Professor, Business Information Technology

Minot State University

Preface

Wargaming is a structured simulation of real-world scenarios, serving as a vital tool for exploring complex events in the future, testing strategies, and identifying vulnerabilities. It enhances preparedness for real-world challenges, whether in military operations or other domains. In the realm of cybersecurity, wargaming plays an even more critical role by simulating cyberattacks, assessing organizational readiness, and improving coordination across departments. These exercises provide a controlled environment where professionals can refine their responses to evolving threats, strengthen overall security postures, and document failures from past strategies.

My background as an infantry soldier, armed security officer, and technical support engineer has given me unique insights into the transition from traditional warfare to cyber warfare. Throughout U.S. history, warfare has been a defining force in shaping the nation, from early conflicts like King Philip's War, the American Revolution, and the Civil War to modern engagements such as the Gulf War, the War in Afghanistan, and the ongoing War on Terror. As history has demonstrated, war evolves alongside technological, political, and economical shifts. The American Civil War marked a significant point in history that changed with the evolution of modern warfare. It introduced industrialized warfare, naval warfare, and total war. The Crimean War during the Industrial Revolution impacted warfare by mass production of weapons, new technologies, and changes in tactical strategies. The Industrial Revolution gave rise to the Information Technology age. Information Technology gave rise to cyber warfare: increased connectivity, malicious software, nation-state involvement, and escalation of cyberattacks. As I compose this book, the Ukraine-

Russia War is ongoing, tensions with China, Iran Proxy War, War in Darfur, Yemen Conflict, Syrian Civil War, and Israeli-Palestinian Conflict continue.

As we navigate this new era of technological advancement, the importance of wargaming strategies cannot be overstated. Understanding, testing, and anticipating cyber threats through strategic simulations will be essential for securing national defense, corporate infrastructures, and global networks. This book explores the critical role of wargaming in cybersecurity and its implications for the future of digital defense.

Introduction

In today's digital landscape, businesses face an ever-increasing range of threats that can disrupt operations, compromise supply chains, and expose sensitive information. This has prompted a renewed focus on the increasing threat of cyber gangs and economic uncertainty to protect critical infrastructure. Other global crises continue to sabotage business functions, negatively affecting financial markets, and compromise trade secrets. While traditional security frameworks provide a basis for protection, they remain largely reactive. Forward-thinking planning strategies are emerging as game-changing solutions that have the potential to provide proactive, strategic, and adaptable protection for business infrastructure.

For a long time, companies have adopted risk management frameworks to help business leaders address potential risks using established models, such as cybersecurity risk management frameworks, ISO 27001 (Sabidi & Zolkipli, 2024), and business continuity plans for unseen disruptions (Efe, 2023). While these frameworks offer some structure, they are often rigid and struggle to adapt when the external environment shifts rapidly. For example, during the 2008 financial crisis, many companies with established risk management frameworks found it challenging to adapt effectively to such rapidly changing economic conditions, and they were forced to evolve in real-time (Settembre-Blundo et al., 2021). Similarly, global supply chain plans were no exception during the COVID-19 pandemic, which revealed the vulnerabilities of traditional resilience models and companies' inability to anticipate disruptions of that magnitude (Phillips et al., 2022). To anticipate disruptions, administrators have to review, document, assess, and use critical thinking skills against past strategies.

Although the past strategies were historically important, they tended to suffer from three fundamental weaknesses. The most common were predictable threats such as hackers, challenging economic conditions, and geopolitical risks. These weaknesses evolved more rapidly than traditional security paradigms could accommodate. Another weakness was a lack of simulation, as many models relied heavily on theory rather than testing against real-world scenarios. The third issue was siloed approaches, where organizations frequently looked at one category of threat at a time, rather than utilizing a unified, integrated defense approach.

Information technology departments and businesses face unique challenges because of external cybersecurity vulnerabilities. High-profile data breaches cost companies in the United States an average of $70 billion annually (Bossone, 2018), negatively affecting the trust of businesses and individuals in data systems. The scale of damage to the IT infrastructure warrants new actions and solutions. In response, President Barack Obama required the Department of Defense to develop a cybersecurity force that included over 6,200 military personnel and business contractors (Winger, 2023). Each group has its role in safeguarding against cybersecurity threats. Earlier in 2009, President Barack Obama had already noted severe growing national security risks in the cyber domain. Building on the momentum, in 2018, President Donald Trump signed H.R. 7327—*the Strengthening and Enhancing Cyber Capabilities by Utilizing Risk Exposure Technology (SECURE) Act* — to further bolster national cyber defenses (Burt, 2023). Historically, presidents have had to implement new actions and solutions for the defense of the United States' cybersecurity infrastructure.

However, the implementation of presidential acts has not fully addressed the unique challenges facing the United States' cybersecurity infrastructure. According to Clapper et al. (2017), former Director of National Intelligence, James Clapper and the 2013 Worldwide Threat Assessment identified cyber threats targeting military networks. The executive order is 13636, "Improving Critical Infrastructure Cybersecurity," directed by the National Institute of Standards and Technology (NIST) to address cyber vulnerabilities. Executive order 13636 is considered one of the most serious and well-crafted national security responses involving the government, interest groups, and private sector representatives (Trim, P., & Lee, Y. I., 2025). The National Defense Authorization Act (NDAA) for Fiscal Year 2019 introduced cybersecurity requirements designed to impede efforts by adversaries from gaining unauthorized access to business contractor networks (Trope, 2020). These requirements include maintaining whitelists or blacklists to approve or reject suppliers, products, or businesses from specific foreign governments. Despite these executive guidelines, acts, and mandates, a significant gap remains, particularly for businesses and information technology departments that still face unresolved vulnerabilities.

Cybersecurity and data breaches continue to be a pressing challenge for organizations. Financially devastating breaches exceeding $650 million and $126 million in losses significantly show a concerning incompetence among some cybersecurity managers (Chen, J. et al., 2022). The financial burden of such incidents calls for stronger measures, particularly in defense contractor environments. Managers, administrators, and supervisors have ethical decisions to consider when dealing with zero-day vulnerabilities, cyberattacks, procedural weaknesses, and the broader implications of doing business with the federal government. Ashraf (2022) stated that on average, a data breach would cost $255 per record, and impacted customers will stop doing business with the organization where the data breach occurred. Data breaches are a recurring issue, and many organizations do not have viable long-term solutions. There could be a solid solution, but when cyberattacks cost businesses millions of dollars, a single fix is rarely sufficient. The economic consequences of data breaches are mounting. Barati and Yankson (2022) noted that the Identity Theft Resource Center reported 1,862 data breaches in 2021, up from 1,108 in 2020. It is a year-by-year rise pointing to both growing threats and weaknesses in data protection practices. Incidents where critical information is stolen, seized, or misused without authorization significantly damage the United States' infrastructure.

Unlike traditional processes, wargaming methods utilize military-grade simulations, real-time scenario testing, and multi-disciplinary thinking to stress-test business infrastructures against known and potential risks (Winger, 2023). Wargaming ideas are built around three primary conditions. The first is dynamic adaptation, where businesses continuously revise and update threat scenarios to remain resilient against evolving risks. The second is cross-functional engagement, where wargaming integrates expertise from teams in cybersecurity, financial risk management, and operational resilience. The third is data-driven decision-making, where decision-makers rely on AI and predictive analytics for making informed decisions based on actionable intelligence rather than guesswork.

In this book, the five wargaming strategies are not simply theoretical concepts untouched by crisis but practical methods continuously tested, examined, and refined to help organizations strengthen their businesses against both expected and unforeseen risks. Each chapter reinforces the focus on strategic preparedness, providing you, the reader, with the methods, ideas, and framework to build informed organizational resilience in an uncertain future.

Figure 1. Wargaming Plan of Action: 5 Essential Strategies

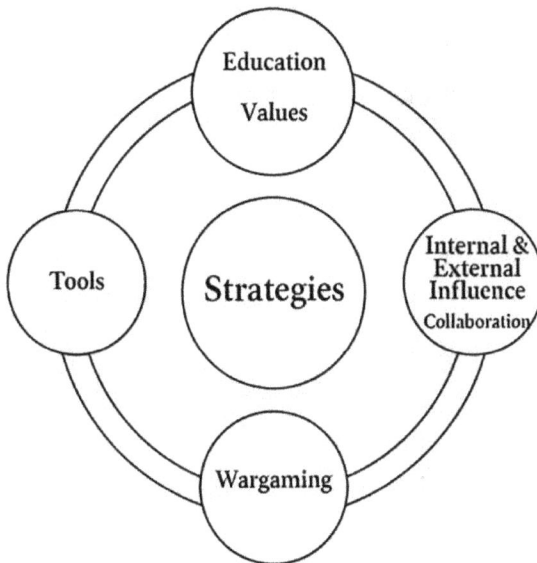

Chapter 1: Cyber Wargaming

Cyber wargaming is a structured, scenario-based simulation designed to model cyber conflicts, attacks, and defensive strategies. It serves as a tool for analyzing decision-making processes, evaluating security postures, and enhancing resilience against cyber threats. By incorporating principles from game theory, military strategy, and cybersecurity, cyber wargaming creates controlled environments where adversarial engagements can be studied to identify vulnerabilities and strengthen incident response capabilities

Rooted in military wargaming traditions, cyber wargaming has evolved to simulate real-world cyber threats in digital domains. By utilizing techniques such as red teaming, blue teaming, organizations can proactively test their defenses. Red teams mimic adversary tactics, while blue teams defend against these simulated attacks. Blue team policy is about revealing vulnerabilities and strengthening security. These methodologies enable organizations to refine cybersecurity strategies, improve threat detection, and enhance overall cyber defense preparedness. Through cyber wargaming, teams can identify potential procedural weaknesses, develop effective software defensive plans, and stay ahead of emerging threats.

To ensure wargaming strategies are effective in protecting business and military infrastructure, collaboration among governments, private industries, cybersecurity experts, and military organizations is essential. National defense agencies, the intelligence community, and cybersecurity firms must prioritize AI-driven simulations, proactive defense strategies, and cyber-physical security. The U.S. Department of Defense (DoD), Cybersecurity and Infrastructure Security Agency (CISA), NATO, and the U.S. Space Force play a critical role in funding, testing, and implementing five network information security wargaming exercises (Liu, L. et al., 2023). These exercises are designed to anticipate and counter threats. These exercises focus on key responsibilities, including ensuring regulatory compliance, protecting critical infrastructure, and developing AI-powered threat detection systems. The artificial intelligence powered systems will anticipate and defend against both cyber and physical attacks.

Private sector companies, particularly those in the automotive, aviation, nuclear energy, and artificial intelligence development sectors, also play a critical role in securing infrastructure. To design and implement the next generation security protocols, such as artificial intelligence-based war simulations and quantum encryption solutions, companies must understand infrastructure strategies. Companies must also effectively uncover and find solutions to general business problems. Tech giants and defense contractors have to work side by side with government agencies to achieve practical business solutions. At the same time, policymakers and regulatory bodies need to draft international laws and ethical frameworks. International laws, policies, and legal frameworks can help to prevent the weaponization of AI and the escalation of cyber physical warfare. Ultimately, success in cyber-physical security comes down to continuous innovation, real-time threat intelligence sharing, and resolving general business problems specific to their industry.

The general business problem is data breaches and cyberattacks on business contractor computer networks. Chua (2021) provided evidence of phishing attacks, ransomware, and cyberattacks that destructively affected the U.S. Department of Health and Human Services (HHS). The lack of security measures to protect against cyberattacks annually on the federal government's information technology infrastructure has a negative impact. Excessive cyberattacks have had a devastating influence on the financial, economic, political, and military systems. In terms of billions of dollars of intellectual property, data assets have been subject to cyber breaches on unarmed and noncombatant networks (Sharma, 2024). The unarmed networks will be described as civilian and noncombatant networks identified by commercial contractor systems.

Another foreign threat and supply chain issue continues with foreign actors engaging in daily data breach infiltration. Ii (2020) provided evidence of Russian nationals' exhaustive cyberattacks against election software contractors. Russian nationals and their companies have posed as legitimate software companies to gather national intelligence. The poor security standards, compromised products, and compliance issues of federal software contractors have led to increased infiltration by foreign actors. This infiltration, which adversely affects contractors' systems, is a reason to sound the alarm and allocate resources to address the issue. As a result of cyberattacks, executives are not surprised that foreign adversaries from China, Russia, and Iran have relentlessly engaged in cyber breaches (Azubuike, 2023). These cyber-breaches target the United States military and business contractor networks. A significant number of unique cyberattacks can focus on a particular agenda for specific financial, economic, and political motives. Exclusive data breaches in specific scenarios can undermine the morale and effectiveness of military personnel, civilian employees, or contractors.

The specific management problem is the implementation of wargaming within the business contractor networks by systems administrators, resulting in data breaches, cyberattacks, and incident-handling processes (Caron, 2021). Cyber wargaming is a remarkable virtual tool used in exclusive behaviors to assess data breaches. The military and business contractor sectors share close relationships, and cyber data breaches have shown adverse effects. Richard Ledgett, former deputy director of the National Security Agency (NSA), Michael Hayden, former CIA director, and James Ellis, US Strategic Command, stated that state actors have the utmost capability of engaging in strategic cyberattacks against the United States' defense infrastructure (Smeets, 2018). The United States' head of state is concerned that cyber weapons have been viewed as special. These special weapons are adversary attacks similar to those in July and August of 2008, when Russia launched a distributed denial-of-service attack against Georgia's network, targeting Georgian news and government websites.

The poor level of compliance between the military and business contractor networks has exposed gaps in cybersecurity management. According to an indictment and a U.S. Securities and Exchange Commission ("SEC") United States versus Iat Hong civil case, three Chinese nationals hacked computers containing information related to high-profile merger-and-acquisition ("M&A") transactions (Trope, 2019). Client trade secrets, financial, healthcare, and law enforcement records across business contractor networks are vulnerable to data breaches from Chinese hackers due to gaps in cybersecurity management. Consequently, a significant organizational issue is the need for systems administrators to implement wargaming within business and occupational contractor networks.

13

Military and business contractor networks have failed to maintain cybersecurity compliance, leaving critical infrastructures vulnerable to cyberattacks, espionage, and data breaches. SolarWinds, an IT firm, was victimized by a supply-chain attack where hackers injected malicious code into their software update system (Cassottana et al., 2023). This created a backdoor to the IT systems of customers, allowing hackers to access confidential information of US firms and government agencies. The attack went undetected for months, highlighting the stealthy nature of such cyber threats. Another example of this is how this breach, along with many other cyber intrusions, has demonstrated an urgent need for a comprehensive cybersecurity strategy that incorporates wargaming simulations. Without robust cybersecurity measures, risks will escalate, intensify, and increase. This is a scenario that is compromising the protection of classified defense contracts and potentially destabilizing the global economy.

The structured wargaming exercises must be incorporated into the cybersecurity frameworks. If brought into contractor networks, wargaming can play the role of simulating real-world cyber threats, stress testing defenses, and practicing personnel in appropriate roles. The users, personnel, and human assets can respond strategically to an actual incident. According to cybersecurity experts like Sarjakivi et al. (2024), the 'whole of nation' response is recommended, where governments and private actors team up in the core of information sharing, attack simulation, and proactive measures for slashing threats. The US Navy's Large-Scale Exercise 2021 involved 25,000 participants across 17 time zones, testing a new warfighting approach that incorporates a Synthetic Environment (SE) (Budning et al., 2022). SE uses technologies like AI, machine learning, and gaming industry tools to create realistic computer simulations. One of its aims is to improve the command and control (C2) component and to isolate vertical or siloed methods of operations. This is where different departments, teams, or systems work independently with little collaboration or integration with others. On the other hand, different cybersecurity approaches can be an afterthought for some organizations. Organizations must consider wargaming as a mandatory practice in defense. Nonetheless, national and corporate security requires immediate action.

The lack of implementation of wargaming strategies or models, including the York Intelligence Red Team Model, demonstrates a significant gap in how the military and business contractor computer networks protect against data breaches. Additionally, red team students often lack real-world experience as business defense administrators and may not have hands-on experience with security assessments. The security assessments are questionable. Penetration testing supports ethical hacking and can be used in wargaming, but wargaming encompasses a large range of techniques and scenarios. Although a flexible ground framework is implemented, it is limited because it is employed in distinct communities, rather than in government contractor networks. Attack tree techniques have not been widely adopted in modeling approaches for business defense contractor networks. More so, cyber defense simulation exercises are not the same as defense system administrators interacting with daily data breaches on critical networks.

Administrators using diminutive wargaming strategies are filling the gap in practice that ineffectively addresses data breaches on business defense contractor networks. Schwarz et al. (2019) mentioned tactical competitions, situation preparation, and enterprise confrontation tournaments that produce lackluster results. The lackluster results showed that critical research is needed to understand wargaming implementation strategies in the United States' information technology infrastructure. Burt (2023) proposed adversarial machine learning environments in cyber warfare settings that utilize game theory in a restricted, simulated system, rather than an actual military defense network. Business defense administrators lack the cybersecurity tools to test, modify, or simulate wargaming scenarios on networks linked to critical infrastructure. When used constructively and appropriately, adversarial machine learning will provide another wargaming technique against data breaches. Therefore, more research should be conducted to understand the implementation of wargaming strategies by business defense systems administrators to gain more knowledge on how to address data breaches and cyber threats.

Wargaming is also described as a future-proof innovation that can assist the business defense industry in incorporating new techniques and procedures against data breaches. The topic of wargaming, defense contractors, and data breaches is relevant to the current literature (Efrony & Shany, 2018). Therefore, there is a need for more studies to explore wargaming strategies that incorporate cybersecurity in the military networks linked with business defense contractors. Business defense field documents, critical materials, and experiences are vital to assist system administrators in discovering new knowledge on sophisticated data breaches against the United States' technology systems. The duties of system administrators involve support, diagnosis, repair, and prevention. They are multi-disciplinary experts who demonstrate technical, administrative, and socio-psychological skills. Most importantly, the duties of business defense system administrators involve solving problems to protect the American infrastructure.

Chapter 2: Significance of Wargaming

The use of wargaming strategies in cybersecurity, critical infrastructure protection, and military defense is the job of policymakers and security professionals. Business leaders, system administrators, and contractors who manage and protect sensitive networks also play a crucial role. Given the growing threats to cybersecurity, including those launched by nation-state hackers and AI-powered cyber-warfare, it is imperative that every stakeholder actively contributes to strengthening wargaming practices to build organizational resilience (Sarjakivi et al., 2024). Wargaming has a long history, but its immersive nature has limit its integration with social scientific advances. In 2019, the International Crisis Wargame demonstrated the viability of experimental wargaming (Schechter et al., 2021). Administrators can use experimental wargaming in military operations, conflict resolutions, and international relations. The Administrator's role in addressing this issue is critical. If security measures remain fragmented and reactive, data breaches, economic losses, and national security risks will continue to plague organizations.

As an IT professional, decision-maker, or security officer, it is your responsibility to integrate wargaming into your organization's cybersecurity framework. Administrators and academics use wargame simulations and formal models to study state interactions that are difficult to observe directly (Oppenheimer, 2024). For wargames to be effective, they must accurately reflect real-world policy options to ensure external validity. As Ahmed and Gaber (2024) note, participating in cyber wargame exercises is highly beneficial for businesses and intelligence contractors, helping them prepare for Advanced Persistent Threats (APTs) and zero-day exploits. Ultimately, it is up to security specialists to determine whether an organization becomes resilient or falls victim to cyber threats. Administrators are responsible for securing networks, safeguarding sensitive data, and anticipating threats through strategic wargaming.

A research-based book is vital because, despite historical efforts over the past five years, cybersecurity attacks have not decreased. In 2015, G-20 leaders and the China-U.S. bilateral joint statement addressed concerns over state-sponsored cyber activities, specifically the use of stolen business, scientific, and technical data for reverse engineering projects. In 2016, the Federal Bureau of Investigation reported a significant data breach affecting 30,000 federal employees, as well as a separate breach compromising the personal data of nearly 300,000 Army National Guard (Mahbod et al., 2019). The lack of strong procedural and technical safeguards to protect service members' information and classified engineering projects has created a vulnerable environment, exposing individuals and national assets to undue harm. Further, in February 2016, major cyberattacks struck the Internal Revenue Service (IRS) networks. That same year, in July, the email accounts of United States Democratic National Committee members were breached, resulting in the release of over 19,000 emails on WikiLeaks (Topor, 2024). The IRS breach had significant consequences: compromised identity, financial data, and personal information, which have been exploited in tax fraud schemes involving virtual currency laundering, terrorist financing, and the illicit trade of personally identifiable information (PII). These attacks have led to increased tax-related identity theft, heightened risk to taxpayer accounts, and significant economic losses. There is ample evidence that, to date, government organizations have not succeeded in eliminating or even significantly reducing data breaches during politically sensitive events and within critical systems like taxation.

Historical efforts to address cybersecurity threats continued under the administration of the 45th president of the United States. In 2017, President Donald Trump's *National Security Strategy* (NSS) promised an "America First" approach, pledging to prevent the theft of sensitive domestic information and protect national interests by using cyber capabilities (Ettinger, 2018). These cyber capabilities were designed to deter sabotage and prevent efforts that undermine the legitimacy of the national process. Further reinforcing this approach, on May 4, 2018, President Trump issued a memorandum that initiated the establishment of a Unified Combatant Command dedicated to coordinating cyberspace operations and defending the Department of Defense's information network (Galbraith, 2019, and Goldsmith, 2022). However, despite these federal efforts, business contractors supporting the government who failed to adopt structured cybersecurity practices, such as wargaming, did not succeed in preventing considerable losses of sensitive information. Industries, including manufacturing, industrial operations, and the broader supply chain, remain frequent targets of cyberattacks by foreign adversaries, specifically Russia, China, North Korea, and Iran. These repeated incursions emphasize the urgency of this research, which seeks to highlight countermeasures strategies, proven wargaming practices, and cyber capabilities tailored for defense system administrators. The research aims to provide practical insights for strengthening the cybersecurity posture of contractor networks linked to national defense infrastructure.

Significant cyberattacks and data breaches continue to pose serious threats to government information networks. These cyber-attacks cause widespread disruption and undermine public trust. While the Federal Cloud Computing Strategy mandates that vendors and government agencies evaluate and approve secure cloud-based solutions for digital infrastructure (Kommidi et al., 2024), this mandate alone has proven insufficient in preventing or mitigating cyber threats. The persistent nature of these breaches reveals critical gaps in the current cybersecurity framework. To address this challenge, implementing wargaming techniques, strategies, and procedures by system administrators is essential. These proactive measures can offer valuable insights into potential vulnerabilities, improve incident response preparedness, and ultimately reduce the impact of cyberattacks on government networks.

State departments and independent agencies of the United States federal government have increasingly faced data breach vulnerabilities due to poor cybersecurity compliance and oversight. The Department of Defense and the Federal Acquisition Regulation mandated the usage of less competitive sole contracts and approval for acquiring products and services (Sanders et al., 2022). These federal mandates attempt to simplify acquisition procedures during cyber incidents involving protected defense information. However, the data breach at the Office of Personnel Management highlighted a serious vulnerability in contractor information systems, ultimately compromising services tied to both government and military networks. In response, the Defense Federal Acquisition Regulation Supplement Case 2016-D025, *Liability Protections When reporting Cyber incidents,* was established to encourage more accountability among contractors, many of whom were previously hesitant or unwilling to report breaches (Woods et al., 2023). This document suggests that effective writing can enable authors to clearly communicate logical arguments to defense system administrators. This provides them with strategies to reduce cyber risks and prevent potential losses of future defense contracts resulting from cybersecurity breaches. In essence, the study proposes that well-written documents can help defense organizations better understand and mitigate cybersecurity threats, ultimately protecting their contracts and reputation.

Military knowledge, training, and technology often leak into the business contractor sector, making it vulnerable to exploitation by foreign agents engaged in espionage. According to Benaroch (2019), notable firms such as QinetiQ, RSA Security LLC (a cybersecurity subcontractor of Lockheed Martin), and Booz Allen Hamilton have been targeted by both foreign and domestic actors in significant data breaches. In some cases, defense contractors did not report these breaches, possibly due to concerns over brand reputation or inadequate cybersecurity capabilities. This investigation presents opportunities to discover unique wargaming strategies used by defense system administrators.

United States citizens must be aware of the unsettling impact that cybersecurity failures within business defense contractor networks have on national security and their personal lives. State-sponsored cyber actors have successfully gained access to staggering volumes of sensitive data, including personally identifiable information of 110 million individuals, 500 million user records, 3 billion sets of credit card data, 50 million user computer logs from retail systems, and even Defense Department travel records (Li, C., 2019; Payton & Claypoole, 2023). According to Trope (2020), current legal frameworks do not require business contractors to investigate, disclose, or report data breaches that compromise sensitive data. It is vital that stakeholders learn and discover how defense system administrators implement cybersecurity procedures to address and reduce vulnerabilities.

The devastating impact of data breaches on military, government, and business contractor networks has warranted the establishment of bug bounty programs as a critical cybersecurity strategy. These programs operate on a signature-based model rooted in adversarial testing to identify valid vulnerabilities while minimizing false positives. Bug bounties are designed to verify vulnerability patches, reduce the recurrence of cyber incidents, and enhance collaborative threat detection. In response to rising cyber threats, the U.S. Congress enacted H.R. 7327 – Strengthening and Enhancing Cyber-capabilities by Utilizing Risk Exposure Technology (SECURE) Act which directed the Department of Homeland Security to establish a bug bounty pilot program (Burt, 2023) While bug bounty programs are a novel concept, they still have the clarity, utility, and efficiency required for widespread effectiveness. To address these gaps, supplementary cybersecurity procedures must be identified to enhance simplicity, competency, and overall resilience. According to Sullivan (2020), the federal government has imposed bans on international business contractors and their vendors if they fail to meet established cybersecurity compliance standards. Former Secretary of Homeland Security Michael Chertoff, Former Director of National Intelligence Admiral Dennis Blair, and the President have emphasized the importance of defending against foreign intelligence entities who engage in cyber intrusions (Young et al., 2023). These breaches frequently target personally identifiable information, intellectual property, and sensitive military research through contractor electronic highways.

Federal cybersecurity standards are insufficient in preventing public cyber incidents, especially during high-profile events. Vedral (2021) acknowledged that cyberattacks have significantly harmed the United States' financial system and national security. North Korea, a long-time adversary of the United States, and China, an emerging superpower, actively pursue American digital assets through sophisticated cyber campaigns. Notably, Chinese hackers have targeted sensitive maritime technologies developed by universities across the United States, Canada, and Asia, particularly those with military applications (Segal, 2020). As a result, public and private academic institutions have become victims in a prolonged campaign to compromise the United States' digital assets. One of the most notable data breach caused by Chinese hackers exposed the personal data of 21.5 million federal employees through the Office of Personnel Management (Chaskes, 2022). Foreign actors continue to compromise, steal and ruin the reputation of United States business assets. Understanding the perceptions and firsthand experiences of defense system administrators is vital to identifying practical vulnerabilities and real-world outcomes of these breaches.

Foreign malicious actors have launched increasingly sophisticated campaigns and cyber operations aimed at compromising the United States' technology infrastructure. One notable example is China's "Three Warfare" strategy, which systematically targets the United States military networks of its defensive, economic, and diplomatic sectors (Neculcea, 2023). The Chinese hackers represent the first structured initiative designed to challenge and undermine U.S. cyber laws, frameworks, and procedures (Smith, 2024). This regime has recognized the value of researching American cybersecurity literature to inform and launch more effective cyberattacks. In January 2018, sophisticated Chinese hackers successfully infiltrated the U.S. Navy contractor, stealing classified weapon development plans (Burt, 2023). These breaches highlighted the persistent lack of robust cybersecurity procedures, technical knowledge, and strategic defenses across American defense networks. Such incidents show that cybersecurity threats are growing worse rather than improving. While existing literature has proposed frameworks and policies post-breach, there is little empirical evidence examining the role of wargaming strategies by defense systems administrators within business contractor networks. This gap shows the need for focused research to explore how wargaming can help improve cybersecurity in this critical sector.

Chapter 3: Wargaming

Business wargaming can be described as a structured exercise involving actionable strategies, capabilities, and behaviors designed to improve a team's understanding of their competitors. Business wargames are a complex practice used to uncover internal vulnerabilities and address external threats (Sisson, 2021). Wargaming mechanisms in business can influence authentic behavior through a feedback loop linked to user actions. Competitors, customers, suppliers, shareholders, investors, and regulators are critical stakeholders in role-playing simulations aimed at understanding practical cyber actions. According to Schwarz (2020), the practice of business wargaming stems from military role-playing simulations involving two groups: the first represents the company's internal strategy, while the second takes on the role of external competitors or adversaries, simulating competitive or hostile actions. Wargaming supports two or three activities played by the two groups. The actions are representative of decades of military-inspired adversarial training adapted for business contexts.

Evaluating, outlining, and explaining relevant frameworks is essential to understanding why the key terms and concepts guide this investigation. According to Johnson and Jordan (2019), an analytical framework is a guide based on an extensive review of qualitative studies, empirical work, and theory. Practical work has its original application on constructs and frameworks. The empirical work, framework, and concepts are composed of one or two research authors. One or two research authors in the wargaming literature can be used to form new frameworks, methods, and policies.

The Open Strategy framework comprises two approaches: inclusiveness and transparency. Developed by Hautz et al. (2017), the framework focuses on the people involved and how information is shared. In contrast, the Cyber Wargaming Framework, which enhances cyber wargaming with a realistic business context, leverages robust cyber defense technology and assesses the business operational impact. Fox et al. (2018) pointed toward the framework developed to focus on securing systems in the financial services sector. However, it will be used more broadly in other critical infrastructure protection sectors. Compared to the Open Strategy framework, the Cyber Wargaming framework has more flexibility in planning, preparing, and conducting cyber business wargames.

The content and process branches are two links of the Open Strategy framework. A similar scenario exists with the inclusiveness, transparency, content, and process branches. According to Schwarz (2020), the Open Strategy Framework has significant practical applications with differing approaches in both analog and digital modes. The analog and digital modes correspond to tabletop and digital simulation of wargaming. The Cyber Wargaming Framework features two distinct elements: its wargaming construction kit and business war games. Fox et al. (2018) showed that cyber wargames have a commercial and business impact. The framework provides a commercial service offering that supports Cisco's Cyber Range, Deloitte's offering, and PwC's Game of Threats.

The Open Strategy and Cyber Wargaming framework have a history of supporting third, minor, and primary parties. The Maelstrom and Fort Meade Experiment (FMX), supported by the Cyber Wargaming Framework, are living laboratory exercise environments for strategies employed by the red team and the theme of advanced adversary activity (Fox et al., 2018). Maelstrom is a board game that incorporates the ATT&CK threat model and framework elements, capturing detailed tactics and techniques within enterprise environments. The Open Strategy framework analog mode supports outside avenues such as workshops, town halls, world cafés, or surveys. Digital modes can host wikis, blogs, and web-based crowdsourcing within the analogous modes of workshops, town halls, and cafes (Schwarz, 2020). Both frameworks apply to similar elements of digital and non-digital business sectors.

Therefore, building on this foundation, the Business Cyber Wargaming Framework introduces eight elements that integrate and expand on traditional wargaming principles. The author developed a new framework by combining sections from the Cyber Wargaming: Framework for Enhancing Cyber Wargaming with Realistic Business Context and the Open Strategy (OS) framework.

Figure 2. Business Cyber Wargaming Framework

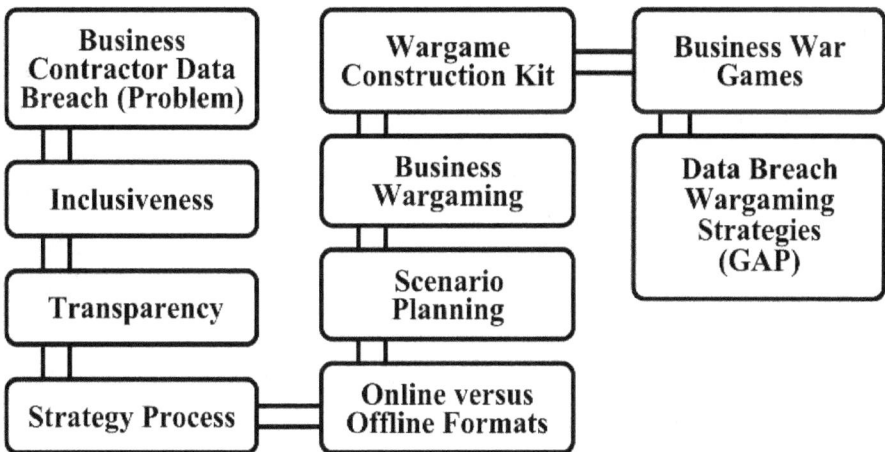

First, inclusiveness is a practice or policy that addresses various subject areas, providing equal access to resources. Coenraad et al. (2020) stated in their research that inclusiveness is related to race and gender, enabling a better understanding of the cybersecurity games' landscape. Heterogeneity, variety, and inclusion ensure talent is accelerating to a highly innovative, resilient operation. Schwarz (2020) stated that the Open Strategy framework, not limited to a wide range of people, includes internal and external strategy consultants and supports openness in terms of transparency. Administrators will have to adopt the idea of participation in the organization's cybersecurity strategy, increasing stakeholder input for decision-making. The information for decision-making supports all sides of the wargaming process. The wargaming process has blue, red, white, and gray teams.

Second, transparency is both a concept and a practice aimed at ensuring accountability, open communication, and clarity. In the context of technology, transparency involves a coordinated vulnerability disclosure process that allows key experts to thoroughly assess a system's security posture (Hughes & Turner, 2023). It can be enhanced by bringing together security researchers from diverse backgrounds to offer new perspectives on data breach vulnerabilities. In qualitative research, transparency enables clear articulation of social processes to all teams and stakeholders involved.

Third, the strategy process focuses on making timely decisions based on verification from multiple data sources. According to Trim, P. R. J., and Lee, Y. (2021), a key part of the strategy process involves two or more entities building strong trust through a collectivist approach to security and a shared set of core values. The strategy process can evolve toward development principles or shift toward implementation strategies. However, it must remain a high-level plan to secure assets, backed by verification from trusted data sources. This plan may also include principles of crowdsourcing. Moradi and Li (2021) point out that different types of attacks significantly impact how adversaries exploit crowdsourcing platforms.

Fourth, online and offline formats represent two states: connected and disconnected. Personalities, motivations, and emotional factors influence how individuals interact with each format differently (Baumann & Utz, 2021). Offline experiences can shape how online strategies are developed and how cyber threats are addressed. The concept of online versus offline formats aligns with two perspectives within the Opening Strategy framework, which is geared toward practical application (Schwarz, 2020). Questions within these frameworks often relate to the Opening Strategy process, scenario planning, or business wargaming. Additionally, effective crowdsourcing should ensure that stakeholders are meaningfully involved in scenario planning or business wargaming conducted in online environments.

Fifth, scenario planning involves making assumptions about technological business planning and addressing specific uncertainties. Van Rooyen et al. (2025) highlighted that scenario planning is a future-oriented technology analysis that includes forecasting, foresight, and road mapping toward complex uncertainties. Its application point toward strategic practices designed to gauge potential outcomes. Administrators facing data breaches must use scenario planning to propose informed responses based on two key phases. According to Sohrabi et al. (2019), scenario planning is a technique used by organizations to develop long-term plans by analyzing trends and identifying potential risks and opportunities. It helps anticipate alternative futures, preparing businesses and policymakers for unexpected, surprising outcomes. These conditions serve as pathways for future transformation and support decision-making analysis. Role-playing scenarios from multiple perspectives add another strategic advantage, allowing for a broader and more effective understanding of cyber threats and potential responses.

Sixth, business wargaming will be described by actionable strategies, capabilities, and operations to improve a team's knowledge about their competitors. Business wargames is a complex practice of uncovering internal vulnerabilities to address external threats (West et al., 2018). Wargaming mechanisms used in business can influence authentic behaviors with a feedback link to user behavior. Competitors, customers, suppliers, shareholders, investors, and regulators are critical stakeholders in a role-playing simulation to understand practical cyber actions. According to Schwarz (2020), the practice of business wargaming stems from military role-playing simulation by two groups. Questions must be targeted toward the first group to represent their firm's strategy. Also, the second group will have to show its role in commercial dealings and competitive moves in business wargaming. Business wargaming supports two or three activities played by the two groups, with actions representative of decades of adversarial situations and vulnerabilities.

Seventh, a wargame construction kit should be viewed as a tool that embodies a system of turn-based strategic battles. Brynen (2020) stated that analog methods have proven flexible, creative, and responsive, even as wargaming evolves into the digital sphere. The analog process of wargaming construction is constructive, helping identify special skills, media acquisition capabilities, and unique training opportunities. Initially, the wargame construction kit is supported by the development of wargame creation skills. These skills encompass a mixture of tasks, including lectures, classroom activities, and independent research projects (Blaschke, 2021). While these skills may not always appear immediately practical, they fill critical gaps in strategic knowledge needed to combat data breaches effectively. The wargame construction kit serves to unify individual administrators' tasks and procedures, providing a cohesive learning environment. There are two main options for implementing the kit: offline or online formats. The traditional tabletop wargame represents elements such as control systems, commands, sensors, forces, and terrain. However, participants' actions can evolve into more advanced, computer-based simulations that incorporate both offline and online cybersecurity components to explore data breach vulnerabilities from multiple angles.

Eighth, business war games are actions designed to pressure-test an existing strategy by simulating opposing teams using standard attack methods. These methods include brute force attacks, DDoS, and zero-day vulnerabilities. In turn, the opposing team may respond with code injection, evasion techniques, and reverse engineering. The value of business war games lies in corporate strategy and the insights they offer into user knowledge (Büchler, 2022). Handling diverse scenarios in business war games supports out-of-the-box thinking and aids in the development of effective data breach policies. Enhancing decision-making capabilities in large groups encourages cooperation and strengthens defensive systems. Business wargaming involves identifying teams, assigning their roles, and defining potential adversaries (Stebbins, 2023). It helps businesses assess how they might function without computers, highlighting vulnerabilities in their current systems. However, formal prerequisites are necessary for business wargaming.

Finally, gaps in wargaming, particularly regarding data breaches, expose a lack of application of game theory, algorithms, and strategic intelligence analysis. Wargames also reveal gaps in management communication and strategic alignment, particularly when analyzing slogans or mission goals. Wargaming uses analytical and behavioral models rather than relying solely on computer algorithms to understand foreign competitors' motivations better. Teams have to immerse themselves in the mindset of their adversaries, living within the competitor's reality, to make accurate predictions about their actual moves. Ultimately, wargaming is about reconstructing a competitor's character to think and act as they would.

Chapter 4: Strategies

The strategies employed by business defense administrators are identified as administrative, deterrent, corrective, and preventive strategy controls. Also, business administrators use and equip themselves with technical, physical, regulatory, and compensating strategy controls. The research, which explores the implementation of wargaming strategies, was conducted in 2023 and involved 10 participants, including defense system administrators, who provided practical information on significant administrative strategies. These participants will be identified in chapters four to eight as P1, P2, P3, P4, P6, P7, P8, P9, and P10. Further details are provided in tables one to three.

Table 1. Participant Job Title, Age, & Region

Participant Number	Job Title	Age	Region
P1	Cyber Planner	32	New York
P2	Systems Administrator	40	Florida
P3	Cyber Engineer	30	Maryland
P4	Systems Administrator	40	California

P5	Cybersecurity Analyst	3 4	New Jersey
P6	IT Specialist	2 8	District of Columbia
P7	Systems Administrator	4 0	California
P8	IT Manager	4 0	California
P9	Information Security Professional	3 9	Texas
P10	IT Manager	2 9	Virginia

Figure 3. Participant Age Information

Age

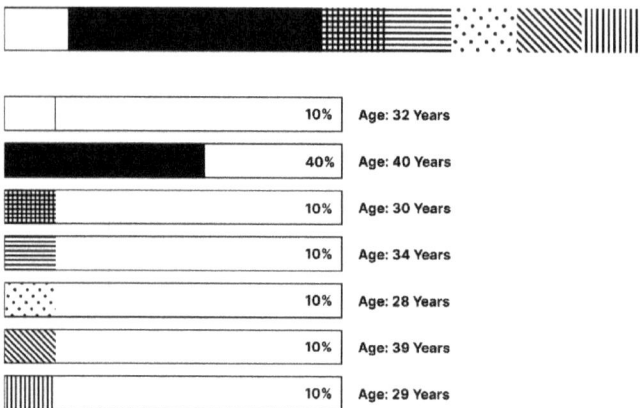

	10%	Age: 32 Years
	40%	Age: 40 Years
	10%	Age: 30 Years
	10%	Age: 34 Years
	10%	Age: 28 Years
	10%	Age: 39 Years
	10%	Age: 29 Years

Administrative strategies implement security culture, policies, and human factors. Administrative strategies play a crucial role in implementing a robust security framework within an organization. These strategies focus on cultivating a strong security culture, where employees are aware of and actively contribute to maintaining security. Additionally, administrative strategies address human factors that can impact security, including the psychological, social, and organizational aspects that influence human behavior and decision-making. Participant P1 offers a unique perspective on various strategies, emphasizing openness, communication, accountability, and data-driven approaches. He also recommends the use of diminishing tactics, actionable strategies, and targeted approaches to address data breaches, affirming, "The only strategy that prevents us from using a wargame construction kit is a kind of tactic we call diminishing tactics." Wargaming is complex and practical because administrators from diverse backgrounds have a unique approach to implementing administrative strategies.

Effective communication strategies in cybersecurity involve carefully planned approaches and techniques. This communication strategy is used to inform stakeholders. Employees, customers, partners, and executives are aware of potential risks, emerging threats, and recommended best practices. Participant P2 acknowledged use of communication strategies that provide detailed analysis for strategy formulations, stating, "Interval readers can actually be dynamic, and also means that we have to be creative about them by having strategy formulations." The policy complies with assistance from the government, which has a remediation strategy for using education and quantitative analysis. Participant P3 noted, "We also have tools that continuously monitor vulnerabilities, which tell us a remediation strategy for securing it." Organizations also utilize advanced tools that provide continuous vulnerability monitoring, enabling administrators to identify potential security risks and develop targeted remediation strategies to mitigate them effectively. These tools empower task forces to proactively address vulnerabilities, ensuring the security and integrity of our systems.

Communication has significantly different techniques that affect strategies compared to previous policies. The evolution of communication techniques has led to a paradigm shift in action plans. Administrators have the option of diverging from traditional approaches and requiring innovative solutions to engage diverse stakeholders skillfully. As a result, contemporary communication strategies must be tailored to accommodate the unique characteristics of various channels, platforms, and audiences. This differs significantly from the one-size-fits-all approach of previous policies. Participant P4 said, "Okay, so we do have different strategies, at least trying to prevent the data breach at first." The current strategy is to obtain a team for a war game strategy to explore an openness strategy. Seminars disseminating information for educating and implementing ideas are a unique decision-making idea. For instance, Participant P7 noted, "Just to make sure everyone is well grounded in any information or organization setup, and also in terms of decision-making ideas, so making sure that all in-house team members are properly informed in case, if any form of law, insecurities or anything when it comes to business decision making." Cybersecurity managers bring balance to the culture of the company. Security Architects play a crucial role in striking a balance between security and usability. By fostering a culture of security awareness and responsibility, they help to create an environment where employees can work efficiently. Participant P8 recalled, "The whole idea of [...] using the old openness helps actually to bring balance within the culture of the company and also [enhance] the whole security system within the company." Modern communication strategies in cybersecurity require tailored approaches for different mediums, networks, & consumers. Cyber defense managers play a vital position in regulating surveillance, accessibility, & collaboration. Innovative solutions, such as war game strategies, seminars, and openness, are used to engage stakeholders, educate teams, and enhance security systems.

Administrators must identify the root of the issue that the business is focusing on. "[...], that is the strategy we use; we try as much as possible to get to the root of the issue, identify where the data breach originated, and determine whether it's on-site or off-site" (Participant P9). In the realm of cybersecurity, accountability is crucial for maintaining trust and ensuring the confidentiality, integrity, and availability of sensitive information. One key aspect of accountability is establishing clear protocols for customer interactions, particularly when it comes to security-related inquiries. Participant P10 shared, "The accountability strategy is only focused on [for instance, where...] a customer wants to interact or ask them specific questions when it comes to security, only authorized personnel [are] allowed to respond." Administrators must identify the root cause of cybersecurity issues to address them properly. Accountability is also pivotal, involving clear protocols for customer interactions. This, ensures only authorized personnel respond to security-related inquiries. Thus, helps maintain trust and protect sensitive information.

Figure 4. Participant Region Information

Region

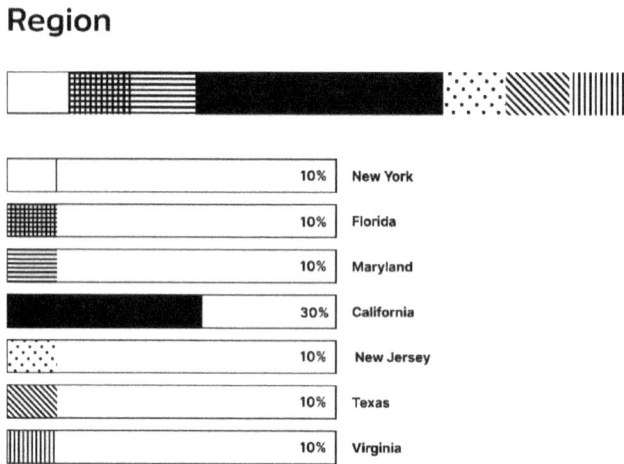

	10%	New York
	10%	Florida
	10%	Maryland
	30%	California
	10%	New Jersey
	10%	Texas
	10%	Virginia

Deterrent controls are a tangible strategy used to reduce a deliberate attack. External influences are tangible objects or people who will cause a cyber incident. According to Al-Janabi et al. (2024), deterrent controls can effectively discourage potential attackers from attempting an attack or prevent users from violating security policies. To achieve this, organizations can implement measures such as warning banners that communicate the consequences of cybercrime, outline legal repercussions for unauthorized activities, and establish robust access controls to limit system access. Participants P1 and P2 confirmed the application of this strategy in their firms. Similarly, P9 added, "And after I did that, that is the strategy we use, we try as much as possible to get to the root of the issue, identify the root of the issue, where the data breach came from, where it is on-site or off-site." Therefore, there is evidence that deterrent controls can identify the location where the data breach came from as an effective strategy. "

Business defense administrators limit their interaction with any external influence with deterrent controls. Participant P10 concluded, "So the thing is that the strategy is based on the overall avoidance policy where the business is being focused on, so obviously, we limit our interaction with more than one type of external influence, whether it is a network or face-to-face interaction."

Corrective controls include incident response plans, backups, and system recovery. Technical corrective controls are a strategy for terminating a process or malware. Administrative corrective control is a strategy for implementing an incident response action plan. The connection between corrective controls and business strategies lies in their shared goal of protecting assets and ensuring continuity. Corrective controls, such as incident response plans, backups, and system recovery, are designed to mitigate the impact of security incidents. In some cases, it helps restore normal operations. Similarly, business strategies often incorporate protective measures to safeguard against threats and vulnerabilities. Participant P2 confirmed this, "In terms of business strategies, we have seen strategies called a common form of protection. You can see how to restrict access [...hence improve] general security. You can see limits in demand." Participant P4 stated, "So, which, at times, can be very damaging when you lose data, so we try to make backups more frequently." Participant P7 shared, "What is the strategy of actions using making timely decisions? I thought there was one, like earlier mentioned, of having a backup plan." Participant P8 mentioned, "Yeah, it is just skilled creation of incident responses." By linking corrective controls to business strategies, organizations can ensure that their security measures align with their overall goals and objectives, ultimately protecting their assets and maintaining business continuity.

Preventive controls are a strategy used with antivirus software, blocking malicious traffic and unauthorized access. They are also a crucial security strategy that employs measures such as antivirus software, firewalls, and access controls to block malicious activity and unauthorized access. By implementing these controls, organizations can significantly reduce the risk of security breaches and cyber threats. Participant P8 mentioned, "Okay, I guess the strategy we use [is that...] you are not providing access to the system." Participant P8 also said, "The strategy is prevention before any incident in the sense that you can have a working system that can detect these breaches." Administrative controls use a strategy to separate duties and open access based on need. Participant P9 added, "Yeah, I guess the main strategy is openness. Being able to get all the information you need." Organizations employ preventive controls, such as antivirus software, firewalls, and access controls. It is a crucial security strategy to block malicious activity, reduce security breaches, and protect against cyber threats.

Technical controls can be viewed as a plan of action using automated software tools, access control lists, and technical configuration auditing programs. Technical ecosystems feature screening systems with alarms and notifications. However, technical administrative controls use a streamlined whitelisting of user roles. An engineer can sometimes set up a bridge to apply a different strategic approach. Participant P1 expressed, "Okay, actually, how it does is when we bring up this approach, and [...], it makes a day, you know, the bridge, the bridge setup approach." A feedback system is a comprehensive system that provides vulnerability checking in the quickest way and application whitelisting. Participant P2 recalled, "Somebody saw us over here. We use such strategies to identify investors who have served their wish by identifying and creating an application whitelisting. We could also use restrictions as an administrative privilege, patch operating systems, and multifactor authentication." Technical controls utilize automated tools, access control lists, and whitelisting. More robust technical controls will employ vulnerability checking, multi-factor authentication with administrative privileges and streamlined user roles for enhancing the security posture.

Table 2. Participants' Experience & Industry

Participant Number	Company size	Experience/Industry
P1	5001-10,000 employees	Information Technology and Services
P2	1001-5000 employees	Information Technology and Services
P3	5001-10,000 employees	Defense & Space
P4	201-1000 employees	Information Technology and Services
P5	201-1000 employees	Computer & Network Security
P6	10,001+ employees	Computer & Network Security
P7	201-1000 employees	Information Technology and Services
P8	1001-5000 employees	Information Technology and Services

P9	1001-5000 employees	Information Technology and Services
P10	201-1000 employees	Computer & Network Security

More than one type of positive tool includes the strategy of implementing practice. "Well, it's about a strategy. I would call it imputation practice, because, you know, each company has their practice, more like a way to respond to different events that will pop up"(P4). The art of simulation can feature a business regulator or control. Participant P6 expressed, "Wargaming business strategies, somewhat like the art of simulation of moves and countermoves, you know, too, I would say stay relevant." Participant P7 said, "Updates in your systems, I think that is one way to prevent any cybersecurity attacks." Participant P9 acknowledged, "And after I did that, that's the strategy we use; we try as much as possible to get to the root of the issue, identify the root of the issue, where the data breach came from, whether it's on-site or off-site." Organizations employ various strategies, including simulation practices like wargaming, regular system updates, and root-cause analysis, to prevent and respond to cybersecurity threats.

Physical controls are a straightforward strategy using physical security keys, access control cards, motion alarm systems, and tangible security control systems. Physical controls provide a tangible layer of security through measures like security keys and motion alarms. Effective implementation often relies on educating team members and providing them with the necessary equipment and training, as highlighted by Participant P7's emphasis on in-house education and resource allocation. Participant P7 recalled "educating in-house team members, providing adequate equipment, or implementing ideas." Participant P9 mentioned, "Yeah, I feel the strategy which I can employ is ensuring that we have all types of equipment, which actually can be needed, just in case of a data breach, you need to prepare beforehand." Physical controls provide a tangible layer of security. The measures used are physical security keys, access control cards, motion alarm systems, and tangible security control systems. These controls are designed to prevent unauthorized access to sensitive areas, equipment, and data. By implementing physical controls, organizations can effectively safeguard their assets and reduce the risk of security breaches.

Regulatory controls are mandatory compliance requirements that offer a strategic advantage to government bodies and administrators when implemented from a legal and authorized standpoint. Participant P6 noted, "But I am just going to take the best guess that the accountability strategy is only focused on, for example, [...]a customer wants to interact or ask them specific questions when it comes to security, only authorized personnel are allowed to respond." Participant P7 stated, "as well as doing all of the strategies here is more like training your employees as well, [...]as informing them on data compliance, and also more like ways to count out a task when there is any form of a data breach of systems." Regulatory controls provide a framework for organizations to adhere to mandatory compliance requirements. By leveraging these controls, government bodies and administrators can exercise their authority to enforce compliance. Effective use of regulatory controls can also help organizations mitigate risks, avoid costly fines and penalties, and maintain a positive reputation.

The strategy of compensating controls involves using countermeasures, impractical implementation, and remediation to address financial issues. Participant P6 shared, "You know, in the case of a data breach or something, we contain the situation by first analyzing it. And then, you know, by providing an incentive to, you know, we are open to our staff, we are like, okay, that is a problem." Participant P9 added, "Yeah, I basically feel that he gets to see the success of how we've improved revenue and how it helps the company chat around better." The strategy of compensating controls involves implementing countermeasures to mitigate risks when primary controls are impractical or impossible to implement. This approach focuses on finding alternative solutions to address financial issues, security vulnerabilities, or operational challenges. By using compensating controls, organizations can effectively manage risks and minimize potential losses.

Table 3. Participants' Gender, Seniority, Education, and Certification

Participant Number	Gender	Seniority	Education	Certification
P1	Male	Manager	Undergraduate degree	Yes
P2	Male	Manager	Postgraduate degree	Yes
P3	Male	Manager	Postgraduate degree	Yes
P4	Male	Manager	Postgraduate degree	Yes
P5	Male	Manager	Postgraduate degree	Yes

P6	Male	Director	Postgraduate degree	Yes
P7	Male	Manager	Postgraduate degree	No
P8	Male	Manager	Postgraduate degree	Yes
P9	Male	Manager	Postgraduate degree	No
P10	Male	Manager	Undergraduate degree	Yes

The ecosystem of telecommunication software features media systems. This ecosystem includes media systems that enable the transmission, processing, and management of various forms of media, such as audio, video conferencing, streaming services, content delivery networks, and multimedia messaging services. By leveraging media systems, organizations can enhance their communication capabilities, improve customer engagement, and increase productivity. Participant P1 said, "So we use software, I mean, our noise, your telecommunication system, so you know, communicate." Participant P3 also mentioned, "Usually, from my point of view, communication comes down to a ticket." The use of software or a telecommunication system to communicate is unclear—the views on communication are primarily related to managing tickets, such as support tickets or issue tracking. Communication is often formalized and structured around specific issues or requests.

Administrative controls support human factors, procedures, awareness, training, and other related events. Clients and coworkers formulate communication strategies based on their openness to different parties. Participant P1 expressed, "An openness strategy, particularly with external servers, allows for free communication within the ecosystem, fostering collaboration and innovation." Open means communication and strategic conversations between clients. Participant P2 noted, "So what [...]we focus on as an entity is ensuring that we have an open means of communication that appears to be the best form of communication between us." Teams and remote meetings hold face-to-face or calls to discuss an open plan of action. Participant P5: "Most of our communication is done via team meetings and remote meetings, whether that is face-to-face or just calls. Some of our events take place in person." While Participant P5 highlighted the importance of team meetings and in-person events for communication, Participant P6 took a more hands-on approach, serving as the primary point of contact and communicator with others. This direct communication approach adopted by Participant P6 is reflected in their description of handling interactions. Participant P6 "I will be the one to, you know, communicate with them, my boss does not go, Oh, I got to talk with these people. This is what is happening." The playbook has formal reports to individuals on a need-to-know basis. Participant P10 was also of the same opinion: "I, along with other key stakeholders, will communicate and create the formal report for these situations," and acknowledged, "Communication is based on only a need-to-know basis."

The deterrent controls discourage external individuals or external entities from causing a data breach attack. Participant P1 explained, "... external forces or external server open strategy allows [...]the free communication in the ecosystem generally". To balance openness with security, companies can limit access to sensitive data. The strategy can consider limiting roles based on user roles, grant temporary access for specific tasks, and restrict access to specific devices. Implementing layered security measures like firewalls, antivirus software and intrusion detection systems can also create a robust defense. A zero-trust architecture that continuously verifies user identities, roles, and permissions can further enhance security. By combining these strategies with network segmentation, automated security monitoring and endpoint protection, defense organizations can protect against external threats. Ultimately, maintaining a secure and open ecosystem.

Physical controls are related to any physical control security system, such as intrusion detection sensors. Participant P1 said, "You know, when we talk about an open system, it is a system whereby you know, different systems, different gadgets." Deterrent controls are security measures designed to discourage external individuals or entities from attempting to breach an organization's data. These controls aim to make the target appear less appealing or more difficult to attack. Examples of deterrent controls include warning banners, security certifications, publicized security policies, monitoring, and incident response capabilities.

Chapter 5: Education (Values)

The overall educational strategy employed by business defense administrators is categorized into administrative, compensating, preventative, technical, and physical control strategies. The access and information subtopics will be discussed further. Administrative controls include practices, procedures, and workplace policies to minimize cyberattacks. The frequency and intensity of such attacks can be reduced through the effective employment of these strategies by business defense administrators. Educational certification training introduces administrators to the RMF framework, NIST guidelines, and documentation policies. As participant P3 noted, "The NIST guidelines, the RMF framework, documentation policies, and procedures guide acceptable use policies and provide clear guidance on how to use the network effectively." The foundational educational framework equips both employees and research teams with a solid understanding of various techniques. This education fosters unique insights and perspectives that enhance knowledge and practical skills. As participant P4 explained, "And this allowed me to come up with a sweet idea, more like, a way I can easily use every technique to tackle what is going on and ensure it does not happen again."

The National Institute of Standards and Technology assessments help business defense administrators conduct administrative analysis. Similarly, Microsoft's assessments provide insights into an organization's administrative capabilities, assisting defense administrators in identifying areas for improvement and making informed decisions. These assessments and tools, such as those offered by Microsoft, can help defense administrators streamline processes, enhance productivity, and gain a deeper understanding of their organization's administrative strengths and weaknesses. Participant P5 shared: "We review the advisories from many different sources, including DNIST, NIST, and the army, which puts out their advisories. So, Microsoft's Patch Tuesdays, we review all these vulnerabilities weekly, if not daily." Educational strategies in this context include brainstorming, encouraging innovative thinking, and educating the team. These approaches enable the team to forecast and develop new solutions for addressing issues that lead to data breaches. Participant P6 noted: "Now, in order to find adversaries, you don't want to be fighting an enemy whose strengths and weaknesses you do not know. It will mean you came to battle on the front unprepared."

Research and surveys serve as instruments for verifying facts and gathering evidence related to cyberattacks. A systematic approach is necessary to collect relevant information, evaluate evidence, and make informed decisions to respond to cyber threats. Participant P8 added, "Yes, if we have a divided decision, gathering relevant information is crucial. For instance, we need clear evidence to choose among options and take action accordingly." Surveys and research materials found in books can also help administrators test their assumptions. Participant P9 acknowledged: "So we have to give a bit of surveys by ourselves, do a bit of research within the company to actually ensure, and also certainly make sure, that we are actually moving in the right direction."

A compensating control can serve as an alternative strategy to meet the requirements of a security measure that may be too difficult to implement. These controls offer viable substitutes when primary measures are impractical, helping organizations effectively manage risk. By leveraging market research, businesses can identify suitable compensating controls that meet security requirements and mitigate potential threats. Challenges such as financial limitations or complex encryption methods may render certain security measures unfeasible. In such cases, compensating controls can provide an alternative approach that addresses these constraints without compromising security. For instance, a compensating control might be implemented where the cost of a direct security solution is prohibitively high, offering a comparable level of protection. Participant P6 emphasized the importance of strategy, stating: "And everything your strategy has to be on point. The business and the market research strategy have to be on point. It has to be, because if you get it wrong there, you just shot yourself in the leg." Participant P9 added, "Yeah, I feel the market research and survey is actually getting to know more about cities, probably charging as much as possible as well to actually see west side opportunities, and also probably getting new ideas." The market research and surveys are essential for gaining localized knowledge, identifying new opportunities, and discovering innovative controls. This information not only aids in understanding the market better but also informs strategic business decisions and generates business ideas.

Preventive controls are strategies designed to reduce and avoid successful threat events before a data breach occurs. One commonly used approach is the need-to-know method, which includes a validation process and clearance protocol to access critical resources. Participant P3 explained, "So there is training that's required for the individual to take and certify that they have taken the training, and then there is a validation process to make sure that they have the appropriate clearance and the need-to-know to access information. So, it is pretty straightforward." Participant P6 added, "Because you do not want information that has not yet been verified, I can say many things." This highlights the importance of verifying the information before sharing it. Unverified information can be unreliable or misleading, and sharing such data can lead to significant problems. Caution must be exercised to avoid spreading unconfirmed claims and to maintain the integrity of information security.

Technical controls refer to software and hardware components used to defend against cyberattacks. Administrators can use test runs and experimentation in wargaming scenarios to assess vulnerabilities and test potential adversaries. Participant P6 observed: "When experimenting with high-risk initiatives, we thoroughly analyze potential hazards. One strategy we use is testing on smaller, controlled projects to mitigate risks". Participant P6 emphasized the importance of such test runs. Similarly, the cybersecurity exhibition highlights securing access points and preventing breaches through robust controls. This proactive approach is crucial in today's digital landscape, where breaches can have severe consequences. "By understanding vulnerabilities, individuals and organizations can better protect themselves." An exhibition focused on cybersecurity topics related to data breaches often explores various means of access, admittance, and admission that could be exploited. These exhibitions examine how such entry points can be compromised and highlight the importance of securing them. By showcasing these vulnerabilities and controls, they aim to educate attendees about the risks, methods of attacks, and strategies of prevention.

To effectively mitigate cybersecurity threats, organizations implement various technical controls that directly interact with specific systems and platforms. These controls manage and regulate interactions with tools such as PowerShell, other sophisticated tools, and unclassified sites. This may involve configuring access controls, firewalls, and other security measures. By establishing such technical safeguards, organizations can better protect their systems and data from unauthorized access or malicious activity. Cybersecurity ecosystems also rely on sophisticated tools to address and resolve physical cyber breaches. Participant P1 emphasized the value of these tools in the broader cybersecurity framework: "So providing access to information in the ecosystem generally has been one of the significant things to our business so far, and we have been using sophisticated tools with the essential help of this advanced technology." PowerShell, a powerful automation and configuration framework developed by Microsoft, was highlighted by Participant P2: "I am sure you must have heard about the likes of PowerShell." This scripting tool is widely used among IT professionals and cybersecurity experts for task automation, vulnerability scanning, and enforcing system policies. Another critical technical control mentioned is the use of firewalls, security devices that monitor and control incoming and outgoing network traffic. Participant P3 shared: "Well, I'll tell you that if something were to happen, we have a firewall in place that will not allow that to happen. As soon as it sees a connection from unrecognized IP sources, it will not allow that connection to come in." Through tools like PowerShell and firewalls, organizations build layered defenses that significantly reduce exposure to cyber threats and improve response capabilities within their cybersecurity ecosystems.

Education plays a crucial role in equipping administrators to apply both apps and offline tools as part of a broader technical control strategy. Their experience highlights a reliance on evidence-based decision-making over speculation or intuition. Participant P6 noted: "On one of the apps, our experience with providing this information is not based on guesswork." This demonstrates a focus on data-driven processes and well-informed practices. Participant P8 shared insights into the challenges of database management: "My company works with getting new information within the database. My company works with acquiring new information within the database. At times, this process can seem deceptively facile, but it requires careful attention to detail [...]." This may allude to the complexity of managing and updating databases, which, although difficult, becomes more manageable and approachable with experience and the right tools. Participant P9 highlighted the importance of accountability in security systems: "Well, in order to do this, we were looking for a security system where the company accounts for everyone." Additionally, Participant P10 emphasized the flexibility and standardization involved in technology implementation: "There is no constraint on the type of tool or system that they want. Everything is asked for enterprise standards, and I set the policies and make sure that all the end users have the best technology to conduct business operations." This highlights the role of leadership in setting high standards and ensuring optimal tools are in place to support secure, efficient workflows. Through education and informed practices, administrators are better positioned to implement adequate technical controls that align with enterprise needs and cybersecurity demands

The implementation of physical controls serves as a foundational strategy to deter cyber breaches. These measures focus on protecting tangible assets such as servers, hardware, and infrastructure by limiting physical access and enforcing on-site security protocols. One key approach involves the use of interim essential tools that have since become accessible to a broader group of users, supporting more dynamic and collaborative data management practices. Participant P2 mentioned, "The strategies we've implemented have yielded useful tools, now available to others, which serve as important sources for system administration and data analysis." This indicates that their organization is actively leveraging a variety of tools and technologies, not just for system administration, but also for strengthening physical control mechanisms.

The strategy of proactively implementing a solution before a threat occurs is called preventative control. These controls place importance on education, teaching, and instruction as essential components of threat deterrence. Proper access to sensitive systems or information requires appropriate clearance and validated credentials. Participant P3 explained, "So if an individual needs to know the clearance level, then they are granted access to information systems. So pretty straightforward. You have to go through some compliance stuff, clearances, and a program needed to obtain the information on the information system." This highlights the structured process involved in ensuring only authorized personnel gain access. Similarly, Participant P8 added, "you need to go to different levels and barricades to be able to access this information." This underscores the multi-layered nature of preventive controls and the importance of rigorous access protocols to safeguard critical information. Preventive controls, therefore, not only reduce the risk of breaches but also reinforce a culture of security awareness and compliance throughout the organization.

Administrative control, including procedures, labels, and training, guides administrators in operating within a secure cybersecurity framework. Participant P2 explained, "If a situation arises that warrants cancellation, also utilize debt restructuring to facilitate automatic country changes and exclude individuals not originally listed in the assets. "This suggests that systems should automatically transfer associated liabilities and restrict unauthorized individuals from accessing assets. A critical aspect of administrative control is proper documentation and the human factor in assigning tasks. Participant P3 emphasized, "If you do not have the right paperwork in place, you are not getting access to it." This underscores the importance of administrative diligence and strict access protocols to prevent data breaches.

Moreover, teamwork plays a vital role in securing systems. Participant P6 shared, "Now, in the online situation, we've got a comparative percentage of the effectiveness and, you know, getting the project done. It saves time, it saves energy." This suggests that evaluating preventive efforts through comparative metrics enhances efficiency and cybersecurity readiness. Face-to-face verification is also seen as necessary in sensitive access scenarios. As Participant P8 acknowledged, "I definitely need to see someone before you can get that information because I just wish I could." This highlights the value of direct human interaction in confirming identity and access rights. Additionally, continuous education is emphasized for decision-makers. Participant P9 said, "As a key decision-maker in the company, I expect people to be willing to collaborate and support our initiatives." This indicates that people in leadership roles are often more inclined to participate in security protocols due to their responsibilities.

Finally, enterprise-wide policies ensure equal access to technology while maintaining strict adherence to standards. As Participant P10 explained, "Everything is asked for enterprise standards. And I set the policies and make sure that all the end users have the best technology to conduct business operations." Administrative controls, therefore, serve as the foundational layer of cybersecurity, aligning procedures, access, and policy with organizational goals.

The information sub-theme focuses on protecting data from unauthorized modification, disruption, destruction, and disclosure. Technical control plays a critical role in this effort and is integral to the cybersecurity ecosystem, which includes web pages, software platforms, and tools. Within their digital ecosystem, organizations often deploy specialized software systems to support secure communication and community engagement. Participant P1 explained, "Generally, in our ecosystem, we organize the kind of system, where we have a software where we communicate, and reach out to every member of the community through that web page." This demonstrates the use of dedicated platforms for structured outreach and interaction. Information from a technical strategy perspective validates datasets through a feedback system built on networking protocols. A two-way feedback loop allows information to flow from one member to another and return to the origin, creating a continuous cycle of validation and refinement. This ensures that shared information remains accurate and relevant. As Participant P2 described, "Should I say a feedback system, where information is channeled from one member in terms of a network form, and also information is channeled back, and such information is used for the input in such a generous system." This collaborative approach fosters data integrity through active member participation and real-time feedback. Further strengthening information security, special systems utilize special codes, auditing software, and network traffic filters to track and verify user activity. However, concerns remain about the sufficiency of these measures. Participant P8 warned, "The security of the information is at risk because each individual has a unique identifier or code that is verified through a tracking system, but there is no additional layer of protection or 'bridge' to prevent unauthorized access."

The administrative controls play a critical role in enforcing team business objectives, particularly in addressing failed recoveries or mitigating the aftermath of cyber incidents. These controls involve setting certain assets as "guaranteed" to ensure their reliability and facilitating the sharing of suitable information among team members to support effective collaboration and informed decision-making. The goal is to effectively manage assets and information within the team to support task completion. As participant P2 acknowledged, "Our approach involves safeguarding assets and verifying that information is up-to-date and shared clearly with the team." A core aspect of administrative procedures is the need for administrators to double-check records and actively filter out suspicious accounts. This process ensures the integrity of resources and helps the team accurately identify compromised assets. Participant P6 emphasized this cautionary approach, stating, "Because it is good to double-check everything because, as I said, you don't want to shoot yourself in the leg. And that is the last thing you want to do for your business and organization." Likewise, Participant P9, the importance of verifying external information sources: "Yes. Like I said before, you need to ascertain the information you're getting from other businesses." In addition, auditing facts, records, and detailed data is recognized as a form of preventative control. It serves to identify discrepancies before they escalate into serious threats. Participant P6 explained, "One of the things that we do is we sit down, and we verify every single one of those pieces of information all over again, in order not to make mistakes." In summary, administrative controls serve as the backbone of cybersecurity governance. Through asset management, internal communication, verification processes, and proactive auditing, these controls uphold business continuity and reduce organizational risk.

Chapter 6: Internal and External Influence Strategy (Collaboration)

Internal influences offer more diverse advantages than external influences due to the direct adoption of means, motives, and opportunities. These internal factors provide greater control, adaptability, and awareness of the security environment. In contrast, external influences tend to establish their rationale based on factors and access obtained from outside networks, often limiting visibility and responsiveness. The study's participants have identified and discussed eight specific strategies under the theme of internal and external influence strategies. The strategies include administrative controls, physical controls, deterrent controls, technical controls, compensating controls, regulatory controls, preventative controls, and detective controls. Each of these controls reflects varying dimensions of internal and external influences on cybersecurity practices, with internal strategies often providing more direct pathways for mitigating risk and reinforcing security posture.

Internal and external influences under administrative principles often involve the delegation of tasks, such as outsourcing to internal departments like sales or marketing teams. However, external vendors differ significantly as they operate outside the organizational structure and often include government entities or third-party businesses. These external parties involved cybersecurity incident responses and bring added layers of oversight compliance. This distinction between internal and external factors highlights the complexities of managing risks and ensuring accountability. While internal teams are embedded within the organizational culture and protocols, they may also be susceptible to internal bias. As Participant P2 observed, "Whereas with other businesses, sometimes internal checks could be influenced to favor one party or the audit." Moreover, government influence plays a critical role in shaping internal administrative processes. Contractors are required to follow specific compliance guidelines set forth by government agencies. Participant P3 noted, "So that is all I have previewed into, I do not have access to what the sponsor or the government person tells the program manager, but from what I see is the program manager puts in the request for the individual to get access." This illustrates a top-down structure where the government sponsors influence internal control decisions, directing program managers who then act as intermediaries. Direct program managers, in turn, request access for individuals. This top-down approach highlights the government's role in determining who gets access, with program managers acting as intermediaries to initiate access to individuals. Such dynamics illustrate how external oversight, particularly from the government, can shape internal security protocols and administrative decision-making.

Internal team members can sometimes disrupt cyber hygiene efforts within organizations. This highlights the importance of educating and aligning team members with cybersecurity best practices to mitigate potential risks. Participant P4 noted, "So this is just the basic stuff that I do to educate my team about investments, talking about disruptive developers in the meeting," highlighting the importance of proactive communication and internal awareness. Insider threats differ from external threats in that internal actors may already have privileged access to systems, making them potentially more damaging. As such, cybersecurity strategies must account for both types of threats. Participant P5 explained, "So in all of our wargaming and theoretical events, we consider whether it's going to be an insider threat that has full access to our systems, or it's an outsider who will be breaking in right from the start." This dual-layered approach helps organizations prepare for a broad spectrum of threat scenarios. When tasks are outsourced to independent organizations or teams, new risks and challenges may arise, particularly concerning oversight and accountability. Participant P6 questioned, "Now, what do we do when we outsource a job, or project, or collaborate on a project?" Such considerations are vital in maintaining control while leveraging external support.

External vendors are often brought in to support organizations in addressing significant cyber threats. These collaborations allow companies to access specialized expertise and advanced technologies, helping navigate complex challenges. Participant P7 reflected, "So this helped in deciding the kind of you know, best or proper solutions or systems we procured in the market space in this phase as well is also helping, you know, connecting with other external, external vendors, as well as making like, should I see a relationship that would help highlight the affairs of your organization's and your long, long term run." The role of internal departments, such as sales and marketing, is also distinct. Unlike technical or administrative teams, their primary function involves interacting with external parties. Participant P10 stated, "And that does not necessarily apply to the sales and marketing team because that's your whole job to interact with external folks." Ultimately, understanding the interplay between internal and external influences is crucial for developing effective strategies. This involves balancing internal risk management with the benefits of external collaboration, ensuring both operational integrity and strategic growth.

The effectiveness of physical controls is influenced by whether they are managed internally by individuals or externally by automated systems. System-managed physical controls, such as electronic locks or biometric scanners, automatically grant access based on predefined criteria, ensuring consistency and minimizing human error. In contrast, individual-managed controls, like manual check-ins or key-based locks, require human intervention, which introduces the potential for subjectivity or errors. As Participant P1 shared, "When a designated individual or system is responsible for collecting and recording specific information, they will typically compile data up to a certain point in time or a predetermined threshold."

Deterrent controls, such as physical barriers, surveillance cameras, and security personnel, primarily serve to counter external threats. These measures create a visible security presence, which can discourage unauthorized access attempts and protect organizational assets. Participant P1 acknowledged this role, stating, "I'm certainly open to a strategy with external influences," referencing the importance of such external deterrents in reinforcing physical security. Government contracts and regulatory frameworks also exert considerable external influence on access decisions. These mandates often stipulate conditions such as clearance levels or project-specific access rights. Participant P3 also noted, "So from an external influence, usually you get access because the government has given you a project or deliverable or a task," highlighting how access is often conditional upon fulfilling external requirements. Additionally, threat simulations frequently include scenarios in which outsiders attempt to breach systems, often using stolen credentials or exploiting weak access points.

As Participant P5 shared, "it's an outsider who is going to be breaking in right from the start, whether they know passwords or not," indicating that organizations must be prepared for a wide range of external threats that test the robustness of their physical and deterrent controls. In sum, internal and external influences shape how physical and deterrent controls are managed. Recognizing the roles of individuals, systems, and external regulatory forces is critical for building resilient security infrastructures. These roles help to address both human and technological vulnerabilities.

External partners often share a strong commitment to achieving project success and upholding the core values of the organization they support. These collaborators are not merely service providers but active contributors to strategic outcomes. Participant P6 highlighted this shared commitment, stating, "We need to factor in external influences that align with our core values, ensuring we're all working towards the same goal – delivering this project with the same level of passion and dedication." This emphasizes the mutual investment and unified focus between internal teams and external stakeholders.

In addition to aligning with values, external partners often play a critical role in identifying and implementing the most effective solutions available in the market. Participant P7 explained, "We're identifying the most effective solutions and systems available in the market during this phase, which also facilitates connections with external vendors, ultimately supporting our procurement decisions." This statement illustrates how collaboration with external vendors is both strategic and essential to innovation and informed decision-making. However, the degree of interaction with external partners may vary depending on the nature of the task. Participant P10 noted, "We don't interact with our external influencers as much, but only if they need specific information." This suggests that while external engagement is valued, it is often reserved for targeted situations, reflecting a deliberate and purpose-driven approach to collaboration. Overall, external partners are more than peripheral contributors; they are aligned with stakeholders whose efforts and values resonate with organizational objectives. Their involvement is structured, intentional, and instrumental in operational success and strategic alignment.

Government regulations, particularly from the Department of Defense (DoD), play a central role in shaping administrative strategies within organizations that support military operations. These strategies often adhere strictly to DoD requirements and standards for information systems. This demonstrates the significant influence of government expectations on internal operations. Participant P3 acknowledged, stating, "What I mentioned earlier is essentially following the DoD defense requirements for setting up any environment for contractors, which serve as hardening guides for your information systems." Similarly, Participant P5 reinforced this connection, noting, "So we provide various systems to the army in the Department of Defense." These statements illustrate the alignment between internal administrative practices and military ambitions, emphasizing a structured, secure approach driven by government standards.

In terms of external threats, Distributed Denial of Service (DDoS) attacks remain a prevalent concern. Organizations deploy a variety of technical control strategies, such as firewalls and honeypots, to mitigate these threats efficiently. Firewalls serve as the first line of defense, enabling systems to detect and block unauthorized access attempts. Participant P3 remarked, "Well, I'll tell you that if something were to happen, we have a firewall in place that will not allow that to happen," emphasizing the system's preparedness to respond to intrusions. In addition to firewalls, honeypots act as decoys designed to lure adversaries away from critical infrastructure. These tools not only detect but also help contain potential attacks. Participant P10 explained, "DDoS from adversaries is based on our detection of the firewall we have implemented. We have honeypots also in our environment, which can layer adversaries to compromise those rather than compromise our network." This layered defense strategy reflects a comprehensive and proactive cybersecurity approach.

DDoS attacks significantly affect networks by overwhelming systems and potentially taking them offline. These disruptions directly impact physical controls by rendering affected systems inaccessible. In anticipation of such events, teams develop contingency plans that account for downtime and system recovery. This includes evaluating how long it would take to restore affected services or switch to a different network. As Participant P5 explained, "We consider if they can take our systems offline, how long it will take us to recover or connect to a different network." From an administrative standpoint, responses to DDoS incidents are guided by structured protocols. These protocols involve step-by-step implementation of effective action plans to mitigate damage and restore operations efficiently. The process begins with identifying winning formulas, followed by systematic implementation and clear delegation of responsibilities. Participant P6 elaborated on this approach, stating, "So we identified a minor error, and we've outlined a few key steps". Here's the step-by-step implementation of all the relevant processes for the solution. "We'll identify what's working, and then determine how to implement that and delegate various operational tasks."

Deterrent controls play a vital function in defending against DDoS attacks by offering early detection and halting threats at their initial stages. These controls are designed to reduce the likelihood of attacks by triggering key alerts and enabling rapid response. Early detection is essential, as it empowers security teams to act decisively before attacks escalate or lead to data breaches. Participant P6 reflected on earlier challenges in handling DDoS threats, stating, "Initially, we thought our approach to mitigating the DDoS attack was correct, but upon further review, we realized there might have been a better way." This sentiment highlights how early stages may have lacked the precision or effectiveness needed.

Over time, deterrent controls proved their value by intercepting attacks early on. Participant P9 emphasized the importance of early intervention, noting, "Let's stop data breaches before they happen by catching threats early, right at the source, before they can cause any damage." These early alerts play a essential part in enabling the information security team to respond quickly and effectively to potential threats, thereby minimizing the impact. Participant P10 further underscored this point by adding, "And also, we'll receive key alerts that'll help the information security team respond appropriately." Such alerts are indispensable in building a proactive defense posture that prevents minor incidents from becoming major breaches.

Both internal and external influences shape data breaches in unique ways, necessitating technical controls that are both proactive and adaptive. A unique aspect of these controls lies in the use of observability tools, data visualization platforms, and communication software like Slack. Slack, typically known for its team communication capabilities, can also be automated and repurposed as an observability tool to monitor system performance and flag anomalies. As part of their regular monitoring routines, teams integrate these tools to enhance system visibility and ensure real-time awareness of potential issues. This structured approach includes weekly check-ins using platforms like Slack, which contribute to consistent oversight and adaptive incident response strategies. Participant P1 explained, "Remember how we use tools like Slack regularly? We can use similar approaches with other software, too."

Technical controls play a pivotal role in safeguarding systems against data breaches through proactive strategies like access control lists (ACLs) and vulnerability assessments. These measures are instrumental in identifying and addressing potential vulnerabilities before they can be exploited. In this context, it was noted that "a strategy[...] would be able to prevent a further data breach; vulnerability checks could lead to loss of certain parts of the organization." A comprehensive vulnerability management strategy would be able to prevent further data breaches; regular vulnerability checks could help mitigate the risk of significant losses to certain parts of the organization. In response to data breaches, best practices emphasize the use of specialized software to erase compromised hardware and rebuild systems from scratch. Participant P3 explained, "The hard drive that the data breach happened on, for example, right, so then that's being zeroed out through special software, and then the system is being rebuilt from scratch." This clean-slate approach is vital to ensuring that no residual vulnerabilities remain.

Further reinforcing this approach, Participant P4 highlighted the importance of drawing from external expertise: "I mean, people in routine who had experience working with APNS searches for data breach, and a couple of them we do make or wish to try to learn about how other companies are dealing with [it]." Such collaborative learning helps teams refine their response strategies using specialized tools and shared knowledge. Additionally, tools like data visualization software can aid in recovery by mapping system partitions, identifying bad sectors, and ensuring system integrity. As Participant P7 advised, "Take a solution that can help prevent any form of data breach or restore your system," emphasizing the dual role of prevention and recovery in technical control strategies.

Administrative controls addressing data breaches are shaped by both internal and external influences, which support the development of distinct classification systems. These classifications guide the identification and resolution strategies used by research teams. Participant P1 noted the role of observability tools and data analysts in addressing incidents, saying, "Should observability tools alone fix data breaches, or is collaboration with data analysts necessary to resolve issues effectively?" However, despite strategic efforts, pinpointing the source or nature of a breach can still prove challenging. As Participant P2 expressed, "We focus on identifying the root cause of an issue and then providing the most effective and efficient solution." Participant P3 emphasized that "as soon as a breach has been identified, you're literally depending on the severity. So, there are different classifications," indicating that responses vary according to the breach type. These classifications help the research team evaluate existing strategies and inform the selection of appropriate responses. Participant P4 supported this approach, saying, "Staying current with software and doing our due diligence on top products will help our company shine and stay ahead of the game".

Administrators also play a proactive role in preparing for a broad range of potential breaches. As Participant P5 explained, "I don't believe this applies to our use case because we're a tactical system from a business standpoint, we prepare for these types of breaches," reflecting a readiness shaped by business-specific needs and administrative foresight. Administrators play a key-based role in evaluating whether a data breach stems from internal or external sources. The initial response involves a comprehensive analysis of the incident's scope and impact. Participant P6 emphasized this process, stating, "Here's the plan: when a breach happens, we spring into action. First, we evaluate the damage and figure out what's been compromised. From there, we can develop a strategy to contain and resolve the issue." However, there may be gaps in the investigation process. As the same participant, P6, later added, "I feel that the identification of stolen sites wasn't thoroughly investigated." Technical teams also take steps to trace the origin of a breach. Participant P9 acknowledged, "Our process involves auditing source code to pinpoint issues, and our engineers investigate whether the source is internal or external." Following identification, a formal reporting process is essential in ensuring transparency and coordinated response. Participant P10 described this procedure: "... the actors that are part of the incident response playbook will be involved, and notification will be communicated to leadership to all the respective parties, and then a formal report will be shared with all the stakeholders." Such structured communication ensures that key parties are informed and that proper measures are taken based on breach classification and impact.

Deterrent controls strategies encompass a range of preventative measures such as firewalls, source code examination, risk analysis, and job rotation. These controls not only aim to discourage malicious behavior but also enhance the organization's resilience to potential threats. Participant P9 emphasized the importance of source code reviews, noting that their team actively scrutinizes code to identify and address potential vulnerabilities that can be exploited. This proactive approach forms a key part of the organization's vulnerability management strategy. Additionally, structured incident response protocols serve as another form of deterrence by demonstrating the organization's preparedness. Participant P10 highlighted this process, explaining that predefined roles and responsibilities are activated when a security incident occurs. "The notifications are sent to leadership and stakeholders, followed by a formal report," he noted, highlighting the importance of clear communication and documentation. Together, these deterrent controls, from technical reviews to formal response planning, send a strong signal of organizational vigilance. By identifying weaknesses before they're exploited and preparing robust response strategies, organizations can effectively deter potential attackers and mitigate risks.

A compensating controls strategy, if used strategically, can effectively reduce errors and mitigate fraud, especially in cases where primary controls are insufficient or impractical. These controls often require additional oversight to ensure effectiveness. Participant P3 highlighted the importance of such measures in threat mitigation, stating, "And then that way, you eliminate that threat. And then you're working to remediate." An important aspect of compensating controls is the use of incentives to encourage problem-solving and security-focused behavior. Organizations can motivate individuals or teams to address complex challenges by offering rewards such as monetary compensation, career advancement opportunities, or public recognition. Participant P6 emphasized this point, noting, "You know, about a data breach or something, both cases, we contain the situation by first analyzing. And second, by providing an incentive to our staff." When a data breach occurs, the organization prioritizes containment through immediate analysis and action. Incentivizing staff to report vulnerabilities and participate actively in incident response cultivates a culture of vigilance and accountability. By promoting proactive engagement and rewarding contribution from cybersecurity efforts, compensating controls not only fill gaps in the system but also enhance the organization's overall resilience.

Compensating control also extends into the areas of counterattacks, threat identification, and incentivization. These measures support proactive cybersecurity strategies that help organizations respond effectively to data breaches. Participant P7 noted the importance of having predefined procedures in place: "So when there's a data breach, you can actually go back on this map and know exactly what to do to counter this attack." A best practice in such situations is to analyze, identify, and respond swiftly to threats. As Participant P8 put it, "I guess you need to analyze, identify the threats, analyze the threats, identify where a threat is coming from, and then respond." Preparedness also requires financial investment. Participant P9 emphasized, "You need to spend some cash to be prepared for such data breaches, so as not to be caught unaware when it happens." Allocating resources toward robust cybersecurity infrastructure and response plans helps reduce the financial and reputational damage caused by security incidents. By combining counter-attack protocols, threat identification processes, and strategic investment in cybersecurity, compensating controls strengthen an organization's ability to contain breaches and minimize impact.

Implementing security measures within a defined structure is a key physical control strategy used to prevent unauthorized access. This defined strategy may include physical barriers, access control systems, and surveillance mechanisms. These are all essential hardware-based strategies designed to deter, detect, and prevent security breaches. By implementing these physical controls, organizations can protect sensitive assets and data from potential security threats. Participant P1 underscored the importance of such measures in the aftermath of an incident, stating, "Okay, this is war gaming equipment after a data breach vulnerability," highlighting the need to secure critical systems post-breach. Similarly, Participant P9 emphasized the responsibility of engineers in overseeing security regardless of the equipment's location, "I said before, whether it's actually on site or off site, that is to the engineers." The implementation of physical controls, such as access control systems and surveillance cameras, helps prevent unauthorized access to sensitive areas and equipment. In the context of war gaming equipment, physical security measures can protect against tampering or theft. These measures, when part of a comprehensive physical security strategy, help minimize vulnerabilities and reinforce the organization's overall defense posture.

The regulatory controls strategy helps administrators anticipate and adapt to changes in the regulatory environment. By staying proactive, organizations can ensure compliance, reduce vulnerabilities, and align with industry standards. Participant P4 emphasized this proactive approach, stating, "You know, making use of software that is up to date, and also ensuring we did our research team out there looking for the best products, that's going to make our company better." By leveraging up-to-date software, companies can reduce vulnerabilities and ensure compliance with relevant regulations. Participant P4's emphasis on research highlights the importance of staying informed about industry best practices and innovative solutions. A proactive approach to regulatory compliance helps organizations mitigate risks and maintain a competitive edge. Ultimately, effective regulatory controls involve ongoing monitoring and assessment to ensure alignment with evolving compliance environment.

Preventive control is a strategic approach aimed at preventing errors, irregularities, and potential breaches before they happen. This strategy emphasizes early detection and system checks as a foundational step. Participant P1 highlighted this standard procedure, stating, "Before we run any vulnerability, that's the normal process before we check the system." A key component of preventive control is understanding data classification and managing data flow appropriately. Participant P3 discussed challenges related to handling sensitive data, noting, "We don't typically experience data breaches, but we do encounter issues with data spills or leaks from higher to lower classification levels." This reflects the importance of adhering to classification protocols to prevent data leakage. It's an intelligent approach to have the skill to analyze and examine data breaches thoroughly. Participant P8 emphasized the ability to examine threats in depth, saying, "And this will help you, in turn, be able to see through [...] the data breach and also ensure that there is no compromise within the company data." In addition, locating the root cause of potential issues is essential for early intervention. Participant P9 recalled, "Initially, we struggled to identify the root cause of our issues, but eventually, we got to the bottom of it. We pinpointed where the problems were coming from and took action to stop them." Altogether, preventative controls rely on a combination of systematic checks, classification awareness, and analytical capabilities to safeguard data.

Detective controls are strategic mechanisms designed to identify errors, irregularities, and vulnerabilities before they escalate into major security incidents. These controls work by continuously monitoring systems and triggering alerts when abnormal activity is detected. For instance, alarms serve as a frontline defense, notifying relevant personnel of breaches in real time. As Participant P1 explained, "When a breach happens, our system sends out alerts and notifications. We quickly notify our engineers, and they spring into action to come up with a plan to address the issue." Detective controls are handy for identifying external vulnerabilities through automated monitoring tools. Participant P3 highlighted this by noting, "And then we also have tools that continuously monitor vulnerabilities, which tell us a remediation strategy on how to secure it." These tools not only detect potential threats but also suggest appropriate responses, making them valuable for proactive security management.

There is an ongoing debate regarding which detective controls most effectively addresses data breaches. Participant P4 expressed, "I analyze previous data breaches, the tools and software used to mitigate them, and the strategies employed to overcome each incident," emphasizing a retrospective approach to improving security measures. Others, like Participant P6, stress the importance of anticipating issues, "One of the things you don't do is you don't wait till the breach happens. You plan ahead, forecast, you predict." Similarly, Participant P7 advocated for strategic planning and redundancy: "Strategize and then make decisions that can help, you know, increase some of these actions [...]that have not been put in place, as well as implementing systems that can help serve like a backup plan in any form of a data breach or any vulnerability in all our systems." Through timely detection, strategic forecasting, and analysis of past incidents, detective controls play a crucial role in reinforcing an organization's cybersecurity framework.

Internal and external influences can significantly extend into defense contractor competitors, shaping how organizations develop and apply control strategies. In this context, administrative, compensating, deterrent, and preventative controls are used not only for cybersecurity but also as tools in competitive positioning. Understanding the strategies of competitors is vital for maintaining a competitive edge. Participant P1 emphasized the value of intelligence gathering, stating, "For you to win the competition, you definitely need to study to have a [...]good knowledge of the competitors." Ethical awareness also plays a role. Participant P2 highlighted the importance of tracking competitor activities, particularly concerning customer behavior, stating, "We could also be able to watch out for what your competitors are doing and the kind of person who hasn't bought their products." This suggests that market observation not only informs product development but also customer engagement strategies.

Cybersecurity concerns also intersect with geopolitical awareness. Participant P3 mentioned the risks associated with products from foreign entities, noting, "We're cautious about using products made in countries with a history of hacking, like China or Russia. We try to avoid them whenever possible due to security concerns." Such comments reflect broader concerns about supply chain security and foreign cyber threats. Finally, robust research efforts are essential to situational awareness. As Participant P4 noted, "You know, because like I mentioned earlier, we do as a research team to go out there and do research about our current situation and see what is going on with other companies and also try to relate our current problems with other companies." This proactive research helps align internal strategies with industry trends and threat landscapes.

A method of evaluating weaknesses and strengths is the best evaluation against competitors. Participant P6 emphasized that this process should be systematic and data-driven rather than speculative, stating, "Every competitor in the same business in the same project, you ever waited, you're evaluating the weaknesses, your strengths, that's what you're doing, you're not just making guesswork." Such evaluations enable organizations to identify performance gaps, improve internal strategies, and benchmark effectively within their industry. However, a persistent challenge in competitive environments is the lack of trust among companies. This mistrust is often rooted in concerns over intellectual property theft, data breaches, and unfair business practices. These risks hinder collaboration and discourage knowledge sharing, which can limit innovation and industry-wide growth. Participant P7 recognized this issue, stating, "I think that will lead to a breakdown in our business, and also thereby leading to, you know, lack of trust from other competitors." Despite the benefits of competitive benchmarking, many organizations remain cautious in their interactions with others in their sector. This caution is particularly evident in highly sensitive domains such as information security. Participant P10 shared an alternative viewpoint, stating, "Using pro teams' knowledge about the competitors, especially in information security, we don't really care for competitors." This perspective suggests that internal readiness and threat anticipation often take precedence over competitor awareness in cybersecurity contexts. Pro teams' expertise helps organizations anticipate and mitigate potential threats. Rather than solely focusing on competitors, pro teams can tap into the organization's knowledge by enhancing their security posture and ultimately, staying ahead of emerging risks.

Compensating controls play a critical role in how organizations navigate competitive landscapes, particularly in the deployment of market research and the design of complex product infrastructure. Participant P3 explained the importance of structured evaluation frameworks, stating, "I guess, system development life cycle or product lifecycle process where we evaluate different products and services out there, compare them against each other along with our requirements, and do a quantitative analysis to determine which product is good. Through this rigorous comparison and analysis, we can make informed decisions about which products or services best align with our needs and goals." This structured evaluation process complements market research efforts, which provide valuable external insights. Participant P6 highlighted this, noting, "Okay, these are some things we need to do, then there's the market. Now, this market research thing, what they do is, they take a survey of what is working in the world, so it's not internalized."

Deterrent controls, on the other hand, are often used to reveal strategic information to competitors. These controls are designed to prevent espionage, sabotage, or aggressive market tactics from rival firms. As Participant P6 pointed out, "One thing you definitely don't want to do is telegraph your moves to competitors." This implies that maintaining strategic ambiguity is a defensive measure that protects sensitive organizational strategies from being exploited.

Preventive controls provide direct protection against competitive threats, particularly in digital environments. These strategies include measures to block malicious traffic potentially originating from rival entities and the separation of competitor-related activities from sensitive internal systems. Participant P6 emphasized the importance of verification and due diligence in this context: "One of the things that we do, [...]to improve a team's knowledge about the competitors, is we ensure that the verification is solid, we do a double check." By instituting layered controls and verification protocols, organizations can safeguard their digital assets while maintaining situational awareness of competitor activities.

Internal and external influences fundamentally shape an organization's internal culture, strategy, and operational effectiveness. These influences often manifest through administrative, technical, and deterrent control strategies, which govern everything from decision-making protocols to security practices. One significant internal factor is the organization's openness to diverse perspectives. Fostering such a culture allows for inclusive decision-making and richer dialogue. As Participant P1 stated, "Fostering an environment where diverse perspectives are valued can be transformative. For instance, organizations that empower individuals to share their unique insights and opinions can harness the power of varied orientations, leading to richer discussions and more informed decision-making." Education and knowledge sharing also play a role in shaping internal culture. For instance, Participant P4 shared, "So this is just the basic stuff I do to educate my team about investment, talking about disruptive developers at the meeting." This highlights how team leaders actively introduce forward-thinking concepts, like disruptive investments, to stimulate awareness and adaptability within the organization.

Administrative controls are further reflected in hierarchical reporting structures. Participant P6 described this structure, stating, "So I report directly to my boss, and my boss reports directly to his boss now, which is worse because it's kind of like a branch." While such chains of command can help maintain order, they can also slow down decision-making and create bottlenecks if not managed effectively. At the apex of internal influence lies the board of directors. Their decisions, vision, and leadership style significantly affect the organization's direction. The board plays a central role in shaping internal dynamics, strategic goals, and overall performance. By understanding the board's role and influence, one can gain valuable insights into the organization's values, priorities, and potential future trajectory. Participant P9 highlighted their impact, saying, "This offered the highest board of directors to invest in more data security in the company." Such decisions from the top trickle down through all departments, influencing priorities like cybersecurity investment and risk management.

Technical controls are vital in safeguarding organizational systems and data. These controls often employ open system plans to enhance both security and operational flexibility. Open systems allow for seamless integration with various tools and platforms, allowing for more effective threat detection and response. As Participant P1 explained, "You know, when we talk about an open system, it's a system whereby you know, different systems, different gadgets." On the other hand, deterrent controls are typically designed to discourage malicious activity by signaling strong postures. Interestingly, in some cases, deterrent strategies incorporate external influences within internal organizational structures. These external actors may bring fresh perspectives or serve advisory roles, helping the organization stay alert to emerging threats and geopolitical risks. Participant P1 noted this approach, saying, "In our organization, we have a person dedicated to monitoring external influences that could impact our company. They keep an eye on potential threats and help us stay ahead." This deliberate integration of external oversight can enhance deterrence by introducing checks that transcend internal assumptions. Together, technical and deterrent controls create a layered security framework. It is a layered framework that combines system integrity with psychological and organizational strategies to strengthen defense mechanisms.

Chapter 7: Tools

Tools in wargaming help address data breaches in network data and detect threat-based activities. These tools extend to data scrambling, web vulnerability scanning, testing, and antivirus software. The participants' interview data identified a main topic called "tools" and subtopics of system, network, data, and ecosystem. Four distinct controls govern tools in wargaming: technical, administrative, physical, and compensating control.

Automated software tools play a significant role in streamlining security processes and enhancing threat detection capabilities. Robust servers provide a secure foundation for data storage and processing, safeguarding against potential breaches. Effective traffic filters help block malicious traffic and prevent unauthorized access to sensitive information. Together, these components enable technical controls to bolster security measures. Technical controls encourage defense business administrators to use automated software tools, servers, and traffic filters as practical data breach tools. Participant P1 recalled, "For customer support, we use Zendesk, and we're also active on X (Twitter), where we respond to customer inquiries and feedback." Zendesk is a cloud-based customer service platform that provides businesses with tools to manage customer interactions, support tickets, and feedback across multiple channels.

The selection of tools and methodologies often hinges on the specific requirements of a project or situation. In many cases, this involves choosing from a range of options, each with its strengths and applications. Depending on the context, specific tools may be more suitable than others, leading to a tailored approach. Participant P1 stated, "Then we use ARIA and ARIA tools, but it usually depends on the kind of condition." ARIA tools enable developers to create more accessible and user-friendly web applications by providing a way to add semantic meaning to dynamic content.

Using password techniques such as Striper Cross reduces costs and provides an additional layer of protection. Implementing robust password techniques is essential for safeguarding digital assets and protecting sensitive information. By leveraging advanced password methods, organizations can significantly reduce the risk of security breaches and associated costs. Effective password strategies, such as those mentioned, offer a valuable layer of defense, as seen in the approach: Using password techniques such as Striper Sross reduces costs and provides another form of remedy.

Participant P2 added, "To prevent further malware attacks, we've implemented security measures like strong password policies and protection against cross-site scripting (XSS) attacks. Tools like DataDog have also helped us monitor and secure our systems." There are a special technical software to help rebuild compromised data after a malicious data breach incident. Passwords are a crucial aspect of digital security, and various techniques can be employed to enhance their strength and effectiveness. Advanced password techniques, such as striper cross or stripe patterns, can provide an additional layer of security. This reduce costs associated with data breaches and offer remedies against scripting attacks. In the event of a breach, specialized technical software can be used to help rebuild compromised data. Additionally, monitoring tools like DataDog can help detect and respond to security incidents. Thus, enabling swift action to mitigate damage.

Participant P3 mentioned, "The hard drive on which the data breach occurred, for example, is being zeroed out through special software, and then the system is rebuilt from scratch." Following the data breach, the affected hard drive is thoroughly sanitized using specialized software to ensure all sensitive information is completely erased. Once the drive is zeroed out, the system is rebuilt from the ground up to prevent any potential lingering vulnerabilities. This meticulous process helps restore the system's integrity and ensures it's secure before being brought back online.

Additionally, the Participant P3 acknowledged, "I tend to focus on the deadline dates because I work with WooCommerce." WooCommerce is open-source e-commerce software for small to large-sized online merchants. It is a flexible and scalable open-source e-commerce software solution designed for small to large-sized online merchants, providing them with the tools and features needed to create, manage, and grow their online stores effectively. It offers a wide range of customization options, extensions, and integrations, allowing businesses to tailor their online stores to their specific needs and requirements. By leveraging WooCommerce, online merchants can efficiently sell products, manage inventory, process payments, and analyze sales data, all within a user-friendly and highly adaptable platform.

Administrators must utilize a variety of versatile tools to identify and resolve backup problems. These tools help to ensure data integrity and availability in the event of system failures or cyber threats. These instruments may include backup software with advanced features, data deduplication and compression technologies. Additional capabilities include monitoring systems that detect potential issues before they become significant problems. By leveraging these tools, administrators can implement robust backup and recovery strategies that protect critical data and minimize downtime. Ultimately strengthening their organization's cybersecurity posture. Participant P4 stated, "What we really need is an all-in-one tool that brings everything together. It'd make it way easier for our teams to collaborate and get stuff done." Tools in a tactical system feature a backup mechanism as an additional security measure. Participant P5 expressed, "There are other TTPs, which are different procedures in place that if something goes wrong with our devices or our software, there are backup mechanisms to fall back on or actual disaster recovery plans."

Administrators have two options: analytical or regulating tools in wargaming scenarios. In wargaming scenarios, administrators have two primary options for analysis and decision-making: analytical tools and regulating tools. Analytical tools provide in-depth examination and forecasting capabilities, enabling administrators to assess complex systems, predict outcomes, and identify potential vulnerabilities. These utilities might include data analytics software, simulation models, and statistical analysis platforms. On the other hand, regulating tools focus on controlling and governing the wargaming environment. This validate that scenarios unfold according to predetermined rules and parameters. These tools, instruments, & resources might include game engines, scenario generators, and rules-based systems.

By leveraging both analytical and regulating tools, administrators can create more realistic, immersive, and effective wargaming experiences. Participant P6 went on to say, "The reason we regulate and control these tools is to minimize damage. By keeping a tight rein on them, we can reduce the risk of major security issues." Participant P7 added, "So, this data, this backup format, or this backup system would help restore all lost information." Dynamic simulation allows administrators to foresee the changes and the course of data breach actions. Participant P8 said, "Yeah, I feel it's the dynamic chip city simulation." Different sets of tools and technologies are used in databases. Participant P10 stated, "So you use multiple tools and technologies that you already have, leverage those, and create scenarios to help the team understand, prepare, and learn."

Tools in administrative controls follow business objectives, risk of remediation, and human factors. In administrative controls, tools are carefully selected and implemented to align with business objectives, mitigate the risk of remediation, and account for human factors. For instance, workflow automation software may be used to streamline processes and enhance efficiency, while compliance tracking equipment ensure adherence to regulatory requirements. This will reduce the risk of non-compliance. Additionally, training and awareness programs are designed to address human factors, such as user behavior and decision-making. These program help to minimize errors and optimize performance. By integrating these tools, organizations can create a robust administrative control framework that supports strategic objectives, manages risk, and promotes a culture of compliance and accountability.

Administrative tools produce alerts to detect anomalies and produce tips on disaster recovery. Participant P3 acknowledges, "And therefore, we can produce alerts, just dashboards that tell us to look into what's happening right from an anomaly detection standpoint." From an administrative perspective, security tools help admins analyze their workflow and schedule backups. For example, security information and event management (SIEM) systems provide real-time monitoring and analytics, enabling admins to identify potential security threats. Risk management professionals can optimize their workflow accordingly. Additionally, backup and disaster recovery tools allow admins to schedule automated backups, ensuring data integrity and availability in case of system failures or data loss. These tools also provide features like data deduplication, compression, and encryption, which enhance the efficiency and security of the backup process.

By leveraging these security gadgets, admins can streamline their workflow, reduce downtime, and ensure business continuity. Participant P4 said, "Trello is another tool we rely on for sharing insights and streamlining our workflow. It's really effective for collaboration and project management." Continuity plans are vital to establishing different continuous recovery plans. Participant P5 shared, "We have different continuity plans in place, depending on the type of event, to see how it affects the mission if we are a tactical system." Participant P6 acknowledged, "So, first is the analytical tool, which involves analyzing the extent of the damage." Participant P10 concluded, "Well, obviously, our security event and incident and event monitoring tool, the SIEM tool that provides us with all the live data that helps us make decisions, and then we also refer to our incident response playbook for the data breach."

The physical controls governing tools include smart devices and security controls. Participant P2 noted, "I've got experience with Samsung, and I work with Xperia devices all the time." Security controls provide additional detection of vulnerabilities and threats. Participant P10 shared, "Additionally, we've security controls in place that provide detection to our incident response team, so they can take quick remediation actions in a situation where we detect anomalous activity."

Tools influence administrators' thinking on how systems behave and the organization's ecosystem. Technical controls in specific

84

systems can be labeled as part of a specific ecosystem, with system restore being one example. Participant P1 noted, "It will bring about great development in an ecosystem so far. So, openness allows, you know, more efficiency and, you know, it makes an ecosystem much better." Participant P3 added, "When we need to modify our system, like adding external access to certain websites, we can make those changes as needed." Participant P4 mentioned, "We prioritize system stability and backups, even if it's just a little data. You never know when you'll need to restore the system, and it's better to be safe than sorry." Administrators view technical controls and system ecosystems as crucial for efficient operations. Ultimately, emphasizing the importance of stability, backups, and adaptability in system management.

System administrative controls extend to quantitative analysis and improving the system through authentic assessments. Information systems must be verified and checked daily. Participant P1 noted, "So, in a nutshell, what I am trying to say is that if the impact it has on the study runnability has been made, then it has a positive impact because the system must have been checked and be intact." Participant P3 also said, "We require our requirements and conduct a quantitative analysis to determine which product is best." Different vulnerabilities negatively impact some organizations' systems' administrative processes of risk assessments. Participant P5 stated, "We do a lot of different vulnerability and risk assessments on our systems prior to going into production." There is an administrative process to improve the system through human factors. Participant P7 shared, "And then the second step is more likely to look for ways to improve our systems."

Two participant responses express physical controls of a proprietary system. Participant P4 recalled, "Backups are key; they help us maintain system integrity and ensure we don't lose everything if something crashes. We can still access what's still working." Participant P5 also noted, "We've had situations where our systems have been taken offline for an extended period, which ultimately affected the mission, requiring the user to read." The compensating controls on the organization system could impact the financial aspects of the product life cycle. Participant P3 expressed, "Yeah, there's a, I guess, system development life cycle or product lifecycle process where we evaluate different products and services out there, compare them against each other."

The networking aspect of tools consists of firewalls and antivirus software. Firewalls and antivirus software are considered technical controls, and administrators tailor their strategies technically. Participant P1 recalled, "Because you have no bad network source, everything will be easy." It's been reported that bugs are present in networking connectivity systems. Participant P2 stated, "We have had issues sometimes with networking connectivity, and a couple of times where we have encountered some sort of bugs in the system, and we had to do debugging." Debugging requires special tools to start and finish the correction process. Participant P3 mentioned, "Well, I'll tell you that if something were to happen, we have a firewall in place that will not allow that to happen."

Firewalls serve as critical procedural tools within technical control strategies, designed to identify and block vulnerabilities, especially those originating from remote hosts. They are essential for filtering traffic, enforcing security policies, and protecting organizational networks from unauthorized access. Participant P4 described the context of using cloud services: "We've got hosting for our email and work files, plus a personal cloud setup for storing and accessing our stuff." In such environments, robust firewall and endpoint protection systems are vital for maintaining security. Organizations often adopt multi-layered solutions to protect their infrastructure. Participant P5 explained, "We use McAfee endpoint security for our cloud systems. We use Palo Alto firewalls, and we use Navios for gathering audit logs." These tools collectively support threat detection, data integrity, and log analysis, forming a comprehensive security posture. The overarching goal, as Participant P10 emphasized, is straightforward yet vital: "Our main goal is to ensure the company's enterprise system and all its networks are secure, too." This underscores how firewalls and related tools are more than just technical safeguards. These tools are a fundamental part of a broader strategy to secure organizational assets against evolving cyber threats.

Technical controls are essential for safeguarding digital assets and ensuring the integrity and availability of data. One significant tool identified in this regard is Datadog, a SaaS monitoring and security platform. As Participant P1 explained, "Okay, so we have many tools we use in this case; we'll use this SaaS tool, DataDog." Such tools are vital in providing real-time visibility into system performance and potential security threats. However, the effectiveness of these tools depends heavily on prompt action. Participant P1 further reflected on a critical administrative challenge: "So the experience so far is that if it's not being attended to immediately sometimes, you know, it makes us lose a lot of data. Yes, we lost a lot, a lot of data." This highlights the importance of a timely response and coordination between technical and administrative teams to mitigate data loss.

Beyond individual tools, technical control strategies also extend to telecommunication systems, which form the backbone of secure and organized communication within an organization. Participant P1 described this aspect of their ecosystem: "Generally, in our ecosystem, we organize the kind of system, the software, we have a software, where we communicate, and, you know, reach out to every member of the community through that web page." This integrated communication infrastructure plays a critical role in disseminating information, coordinating incident responses, and maintaining security awareness across the organization. In summary, technical controls, ranging from monitoring tools like Datadog to secure communication systems, are instrumental in protecting organizational data. Its success relies on proactive management and responsive coordination.

Chapter 8: Recommendations for Wargaming Strategies

The recommendations aim to advance theory, literature, and practical knowledge, linking to the five themes that influence both scholars and practitioners in the business defense industry. Business managers, administrators, and students can gain a deeper understanding of how organizations share similar responsibilities in addressing data breaches. These recommendations contribute to the ongoing development and improvement of strategies for addressing vulnerabilities in cybersecurity data breaches.

The recommendation for theme 1 addresses the inconsistent organizational strategies observed across multiple organizations. Business defense contractors should agree on an industry standard and adopt a generic compliance framework that is flexible enough to allow for tailored strategies addressing specific cybersecurity vulnerabilities. At the same time, it must uphold the core principles of compliance with integrity, availability, and confidentiality across administrative, technical, and physical controls.

Theme 2 recommends effectively implementing best practices using security groups and need-to-know clearance. Administrators must define each specific security group before granting access to digital resources. A baseline minimum of security groups and roles must be established, as this low minimum enables optimum user administration and improves search performance. System administrators can further enhance search functionality by logically naming security groups and classifying need-to-know levels according to different roles. These roles can include admin, contributor, system administrator, guest, and sub-administrator. Permissions should also be managed by checking or unchecking options based on the need to read, write, delete, or access all permissions.

The recommendation for Theme 3 is to determine, classify, and restrict access to internal and external influences. These influences are identified as causes of data breaches. Participants P1, P2, and P4 emphasized record keeping, auditing, and consultation as ways to address poor internal cyber hygiene. Intensive record keeping of all users logging into system resources, along with detailed documentation of administrative, technical, and physical privileges, must be conducted. Additionally, team managers should be consulted on basic cybersecurity hygiene practices, such as regularly changing passwords, updating antivirus software, avoiding blacklisted websites, and refraining from discussing sensitive information. External influences should undergo an auditing process to verify compliance with cybersecurity NIST standards. Experimental scenarios or testing should focus on external distributed denial-of-service attacks. This testing must be conducted using a honeypot or test network to gain insight. The knowledge acquired will equip administrators with vital data and practical methods to respond effectively from future data breaches.

The recommendation for Theme 4 is to utilize a dynamic chip simulation infrastructure enhanced with foreign components of the war gaming experience. It is suggested that businesses consult with vetted computer specialists from countries such as China, Russia, or South Korea. These specialists can simulate real-world scenarios and provide practical recommendations for counterattacks based on war gaming principles. Additionally, they can implement new foreign war gaming strategies using tabletop exercises and other cybersecurity engineering responses.

The recommendation for Theme 5 is to use Kali Linux, an advanced penetration testing Linux distribution used for penetration testing, ethical hacking, and network security assessments. This open-source, security-focused Linux distribution, based on Knoppix, includes more than a dozen tools that can strategically address data breaches. While tools like Slack, Zendesk, PowerShell, and Splunk offer some level of support, but they have limitations when dealing with the frequency and complexity of daily cyberattacks. Therefore, utilizing specialized software with advanced capabilities and tools, such as Kali Linux, is a strategic approach to addressing vulnerabilities.

Wargame Construction Kits is a recommendation aimed at scholars and future research. There should be a reevaluation, review, and update of existing wargame construction kits. Scholars are encouraged to revise the rules, scenarios, maps, and components to reflect the current cybersecurity landscape better. These elements, described as game pieces, theories, rules, and scenarios, must serve a unique purpose. The elements should incorporate international dilemmas, foreign languages, and the influence of social media technologies. Key international dilemmas to include are those involving Taiwan, China, Russia, Iran, North Korea, and India. The toolset must be reexamined to reassess its effectiveness when applied to cyber wargames that include elements such as social media, zero-day exploits, and foreign adversaries. Furthermore, the broader research study should encompass defense contractors based in Europe, Israel, Japan, Singapore, South Korea, Guam, and Iraq.

The development role assisted the researcher in drawing from multiple disciplines. Information technology education contributed to understanding the fundamentals of human development within business organizations. Assurance controls and compliance management aided the investigation by offering insights into various approaches to data protection. The advantages and disadvantages of system application security highlight the challenges decision-makers face in accurately evaluating and comparing multiple protection mechanisms. Senior information technology leaders must navigate diverse governance models, management structures, business engagement processes, and theoretical frameworks.

The implementation of the learning process contributed to the discovery of collaborative leadership roles, which were aligned with business strategy, IT strategy, and ethical policies in strategic implementation. Despite progress, advancements in network security revealed only modest gains in the implementation of protection mechanisms. However, scholarly literature provided significant insights into risk modeling, assessment, and wargaming strategies. The research foundations and activities, while challenging, supported deeper learning. Key learning topics included research design, methodology, population boundaries, and sampling frames. Additionally, the research process involved developing researchable hypotheses and propositions, crafting analysis strategies, and designing effective instrumentation.

Evaluating this scenario introduces fundamental wargaming techniques, concepts, and vulnerabilities related to data breaches. Translating management problems into research purpose statements proved to be a tedious process. However, it guided the research toward relevant theoretical frameworks and the review of various practical models. Critical thinking played a vital role in analyzing practical solutions to cybersecurity challenges. Insights drawn from the existing literature served as a foundation, which could be improved upon or extended to address current business problems more effectively.

This context influences cyber operations, risk management, information security regulations, and the fundamentals of network security. The security architectures used by contractor business organizations often contain design flaws within their data policy frameworks. Business models must align with how data is stored and accounted for existing vulnerabilities. The authorities, roles, and procedures involved in wargaming and cyber operations are complex and require a forward-thinking, adaptive approach. Components, interfaces, strategies, and information systems must be strengthened to support resilient business network architectures. Risk management highlights how both external and internal malicious attacks contribute to data breaches, including intentional and accidental attacks by internal users, as well as frequent daily threats from foreign entities. Environment and structural threats present additional challenges, making them key considerations in developing effective wargaming strategies. Vulnerabilities evolve and may emerge in procedures, locations, software, and hardware. Identifying and managing these risks requires the use of up-to-date vulnerability testing tools.

Information security regulations currently limit administrators in applying cyber defense controls through existing rules and guidelines. These rules and guidelines are often outdated and do not reflect modern wargaming strategies or current responsibilities for countering data breaches. Fundamental concepts such as auditing, evidence collection, and chain of custody protocols must be taught. The concepts and protocols must be consistently reinforced across all stakeholders. Cryptographic elements are notably absent in current wargaming approaches and must be integrated to understand technical vulnerabilities better. The selection of appropriate information security policies, procedures, and access controls should be discussed collaboratively with top stakeholders. Additionally, administrative security policies and operational security tools must be implemented. These policies and tools can be taught across all departments, including external partner companies.

The author provides recommendations to advance theory, literature, and practical knowledge in cybersecurity. It is focused on five themes that influence scholars and practitioners in the business defense industry. These themes address inconsistent organizational strategies, best practices for access control, internal and external influences. The best practices, strategies, and influence extend toward data breaches, dynamic simulation infrastructure, and advanced penetration testing tools like Kali Linux. The recommendations aim to improve strategies for addressing vulnerabilities in cybersecurity data breaches and promote a safe, secure environment. Business defense contractors are encouraged to adopt a generic compliance framework. They are encouraged to define specific security groups and restrict access to sensitive information. The document also suggests utilizing wargaming techniques, such as tabletop exercises and simulation scenarios. Furthermore, it recommends revising existing wargame construction kits to reflect the current cybersecurity landscape better and incorporating international dilemmas, foreign languages, and social media technologies.

Chapter 9: The Future of Wargaming Strategies

The prospect of wargaming leverages human ingenuity, intelligence, and creativity to generate innovative ideas. This fundamentally helps to discover and uncover potential solutions. Wargaming is a valuable tool, strategy, and concept for exploring unpredictable and high-impact events known as Black Swans (Perla, 2022). As technology evolves, so do the threats against the integrity of business and national infrastructures. The advent of cyber-physical systems (CPS), artificial intelligence (AI), and space-based operations has taken the battlefield beyond traditional cybersecurity. Today's threats are not limited to software breaches but also extend to nuclear reactors, automotive systems, avionics, and even space-based assets.

To combat these new threats, wargaming tactics must adapt, fuse AI-driven simulations, incorporate digital twins, and utilize military-grade predictive analytics. Simulation and wargaming, along with emerging technologies like digital twins, can enhance military training, decision-making, and effectiveness (Budning et al., 2022). These wargaming action plans provide tailored feedback, model real-world scenarios, and address skill gaps. This section examines the role wargaming can play in shaping the future of security across major sectors and how it can do so in a way that builds resilience in a world of unprecedented technological change.

Wargaming further leverage human creativity to generate innovative solutions for unpredictable threats. It adapts to evolving tech threats by incorporating AI-driven simulations, digital twins, and

95

predictive analytics. Tactical simulation enhances military training, decision-making, and effectiveness by modeling real-world scenarios. Thus, its role is crucial in shaping future security across sectors and building resilience amid rapid technological change.

Chapter 10: Cyber Physical System Security

Cyber-physical systems (CPS) represent the integration of computational processes with physical operations, forming the backbone of critical infrastructure and industrial control. Securing CPS poses significant challenges due to their complexity, real-time requirements, and the convergence of virtual and physical vulnerabilities. This chapter explores these security challenges and the unique risks associated with CPS environments. Wargaming is highlighted as a vital methodology for testing defenses, identifying weaknesses, and improving response strategies in simulated attack scenarios. The future for leveraging wargaming to enhance the resilience and security of cyber-physical systems is warranted.

Cyber-Physical Systems (CPS)

Cyber-Physical Systems (CPS) are vulnerable to security threats due to their increased attack surface, unknown variables, and common exploits. Traditional security techniques are ineffective against complex attacks, prompting researchers to develop new methods (Sheikh et al., 2023). CPS implements a machine learning-based intelligent attack detection strategy using adversarial learning. This approach is evaluated using Random Forest, Artificial Neural Network, and Long Short-Term Memory on a dataset. This demonstration is effective in detecting attacks and ensuring CPS resilience. Hybrid systems are behind the operation of smart factories, intelligent power grids, autonomous transportation, and even medical life support systems. However, as CPS become more automated and networked, they are also subjected to an increase in surface attacks. As a result, giving businesses and infrastructure opportunities for cyberattacks that can cause physical damage.

The prevalent attacks, intrusion methods, and threats facing IIoT-edge computing environments include denial-of-service (DoS), ransomware, malware, and man-in-the-middle (MITM) attacks. These assults can have a significant impact on industrial operations, emphasizing the need for advanced security mechanisms. A taxonomy of primary security mechanisms for CPS intersect within IIoT-edge computing, including machine learning, federated learning, blockchain, deep learning, encryption, and digital twins. Researchers emphasize the importance of integrating these advanced security technologies to enhance the security and real-time data protection of CPS (Zhukabayeva et al., 2025). Ultimately, potential future research directions aim to advance cybersecurity in this rapidly evolving domain, providing a foundation for future research and development in IIoT-edge computing security.

Cyber-Physical Systems (CPS) face significant security threats due to their increased risk profile and complexity, which can render traditional security techniques ineffective. Researchers are developing new methods, such as machine learning-based attack detection strategies, to ensure the resilience of CPS. Autonomous systems are critical, as attacks can cause physical damage and disrupt industrial operations. Integrating technologies like machine learning, blockchain, and digital twins can enhance CPS security.

Challenges in Securing Cyber-Physical Systems

The recent advancements in exoskeleton technology, including both passive and active types, are significant to securing cyber-physical systems. The sensors, actuators, mechanisms, design, and applications of body augmentation system highlight their use in various industries such as healthcare, manufacturing, and the military. The field of exoskeleton technology presents communication protocols, challenges, and research opportunities (Preethichandra et al., 2024). The importance of developing advanced sensors, actuators, and control systems to enhance the performance and safety of exoskeletons in cyber-physical systems is strategic. By addressing the current limitations and challenges in wearable robotics, researchers and developers can create more effective and efficient solutions for various applications. Ultimately improving human capabilities and safety.

Unlike traditional cybersecurity, CPS security needs to take into consideration both IT (software) security and OT (physical infrastructure) security. The rise of smart cities has led to increased vulnerability to cyberattacks due to the vast amounts of personal and public data generated. Traditional username and password authentication methods are no longer sufficient. A significant number of multi-factor authentication (MFA) solutions have been proposed as a solution. A new concept, "BAuth-ZKP," is introduced, which uses blockchain-based MFA with zero-knowledge proof authentication to secure smart city transactions (Ahmad et al., 2023).

Legacy systems often lack encryption, authentication, and intrusion detection mechanisms, making them prime targets for cybercriminals. Researchers have proposed a multi-layered co-design approach to secure cyber-physical systems by integrating control-theoretic methods and cybersecurity techniques (Fagiolini et al., 2024). Trusted Computing Platforms (TCP) and swarm attestation techniques can ensure software integrity. The proposed method stands out for its ability to describe complex, multi-agent systems and automatically generate code for local monitoring and consensus processes.

Ensuring the security of smart electrical power grids requires effective defense strategies against hacking attempts. To achieve this, administrators must develop a deep Q-learning-based stochastic zero-sum Nash strategy solution (Moradi et al., 2022). This approach overcomes the limitations of conventional Q-learning methods. The framework utilizes reinforcement learning to solve the stochastic game problem in large power-grid systems. This framework has the potential to significantly enhance the security of innovative grid systems, which are critical infrastructure vulnerable to random failures and hostile intrusions.

Advancements in exo-tech are pivotal for securing cyber-physical systems, with applications in healthcare, manufacturing, and military industries. However, CPS security requires consideration of both IT and OT security due to increased vulnerability to data breaches. New authentication methods, such as blockchain-based multi-factor authentication (MFA) with zero-knowledge proof, are being developed to secure smart city transactions. Effective defense strategies, like deep Q-learning-based stochastic zero-sum Nash strategy solutions, are also being explored to protect critical infrastructure.

The Future of Wargaming for CPS Security

Advanced wargaming strategies are decisive for safeguarding Critical Public Services (CPS) against emerging threats. There is an integrated approach to modeling, simulation, and operational planning. Artificial intelligence (AI) can contribute to each of these areas (Davis & Bracken, 2025). By combining operational planning and AI, researchers and policymakers can create more effective decision-making tools. The importance of clear explanation and cognitive modeling in AI is emphasized, cautioning against over-reliance on machine learning. It is suggested that crisis simulation and AI should be used to anticipate possibilities and explore different scenarios, rather than making unreliable predictions.

Nuclear power plants in the US generate 20% of the country's electricity. Nuclear facilities' capacity needs to more than double by 2050 to meet energy demands. To achieve this, cutting-edge technologies like AI and machine learning can increase automation. According to Hall et al. (2024), human-centered AI (HCAI) combines AI with human-centered design to create efficient and reliable systems. A good example is a nuclear reactor cooling system that can be monitored by AI-driven models that can detect unusual temperature spikes, indicating a potential cyber-physical attack.

A low-cost, cross-platform SCARA robot digital-twin simulation system can be use to address the high cost and complexity of testing robot algorithms on autonomous devices. The system establishes a 5D architecture, classifies data, and integrates functions to enable motion trajectory calculation and data storage for a virtual-reality robot (Zhang et al., 2024). Experimental results show that the system allows users to control data communication and synchronous motion. The virtual robot simulation system accurately simulates the physical robot's operating state and spatial environment. This enables safe and convenient testing of robot algorithms. The tested algorithms can be successfully applied to automated platforms, reproducing the operating results of the virtual system. For example, a smart grid operator could simulate a ransomware attack on an energy distribution network to test how automated responses could curtail the damage. Another scenario is when unmanned aerial vehicles (UAVs, or "blue team") use a command and control (C2) network to track enemy agents ("red team"). This scenario models information sharing and awareness using entropy calculations based on agent states (operational level, team, and location) (Tran et al., 2015). In some cases, as trust in information increases, the blue team's understanding of the red team's actions improves. Another wargame example is where hackers try to hack remotely into robotic assembly lines of a manufacturing plant's cybersecurity team.

Advanced wargaming strategies are essential for protecting Cyber-Physical Systems (CPS) from emerging threats. Artificial intelligence can analyze large CPS datasets to identify anomalies and potential attack patterns. This helps proactive defense measures. Digital twin simulations, such as those used for robots or energy distribution networks, facilitate the safe testing of algorithms and responses. Ultimately, wargaming scenarios, involving red and blue teams, can simulate real-world attacks and help develop effective mitigation strategies.

Chapter 11: Wargaming Nuclear Security

Nuclear reactors are critical components of national infrastructure, requiring robust security measures to mitigate both physical and cyber threats. The complexity of these systems presents significant challenges in identifying and addressing potential vulnerabilities. Wargaming has emerged as an essential methodology for analyzing security risks and testing response strategies in a controlled, simulated environment. This chapter examines the specific challenges faced by nuclear reactors and explores how wargaming techniques can enhance preparedness. Furthermore, it considers the future directions for integrating operational preparedness into nuclear security frameworks to support effective risk management.

Security Nuclear Reactors

Among the most high-risk critical infrastructures are nuclear reactors that supply energy, medical isotopes, and national security capabilities. As nuclear power plants increasingly digitize, they incorporate Industrial Control Systems (ICS), Supervisory Control and Data Acquisition (SCADA) networks. There are a few nuclear power plants that also incorporate internet-connected sensors. A significant number of nuclear power plants are becoming vulnerable to cyberattacks. Critical infrastructure (CI) is vulnerable to cyberattacks, posing a significant threat to modern society. CI includes physical and cyber assets, systems, and networks across various sectors, such as energy, healthcare, and transportation. A disruption in one sector can have a cascading effect on others. Furthermore, the security of nuclear reactors is also a pressing concern, as cyberattacks could have devastating consequences. The Stuxnet worm, which targets industrial control systems, including those used in nuclear power plants, is highlighted as a significant threat, having infected over 30,000 industrial computer systems in Iran in 2010, and posing a risk to critical infrastructure worldwide (Riggs et al., 2023). Moreover, with state-sponsored cyber warfare and terrorist threats escalating, there is no greater necessity than advanced wargaming strategies to secure nuclear reactors.

The platform enabled quantitative analysis and causal inference, addressing the limitations of traditional wargaming. Wargaming researchers developed an experimental wargaming platform called SIGNAL to study the impact of tailored nuclear capabilities on the nuclear threshold (Reddie & Goldblum, 2023). Their findings suggested that tailored nuclear capabilities may not increase the likelihood of crossing the nuclear threshold. Additionally, low-yield nuclear weapons may serve as a substitute for high-yield ones. The study demonstrated the effectiveness of experimental wargaming in creating an immersive environment for quantitative social science research and providing valuable insights for academics and policymakers into the future of wargaming.

Nuclear reactors are increasingly vulnerable to cyberattacks due to digitization and interconnectedness, demonstrating impactful threats to countries. The Stuxnet worm highlights the potential for devastating digital assaults on industrial control systems. Advanced risk management techniques are necessary to secure nuclear reactors. Experimental wargaming platforms, like SIGNAL, can provide valuable insights into nuclear deterrence and crisis decision-making.

Challenges for Nuclear Reactors

One challenge is the fact that many nuclear plants have outdated software and legacy systems. According to Boring et al. (2023), as advanced nuclear reactors are developed, there is a need to clarify the role of digital and automation technologies in control systems. However, the outdated infrastructure in existing plants can hinder the seamless integration of new technologies. There are compatibility issues, security vulnerabilities, and operational inefficiencies. To address this challenge, nuclear plants may need to invest in upgrading their legacy systems, developing new digital solutions, and ensuring trained personnel. New reactors, AI/ML technologies, and human-centered AI (HCAI) bring both benefits and challenges (Hall et al., 2024). Most of these systems do not have encryption, authentication mechanisms, or the latest intrusion detection systems. Information technology specialists have concluded that it is an easy prey for cyber adversaries. The lack of robust security measures in these systems makes them vulnerable to malicious attacks. This may allow adversaries to exploit weaknesses and gain unauthorized access. Implementing encryption, authentication mechanisms, and intrusion detection systems is imperative to protect these systems from capable digital assualts.

Researchers explored the role of cyber operations in crisis escalation through operational research games, finding that having cyber response options reduces escalatory behavior. The absence of a set of cyber responses leads to more conventional and provocative actions. One notable example of cyber operations is the Stuxnet worm, widely believed to be a joint US-Israeli operation. Stuxnet was designed to target industrial control systems, specifically those used in Iran's nuclear program (Jensen et al., 2024). It was first detected in 2010 and is considered one of the most sophisticated malware attacks in history.

It is important to examine advanced persistent threats. In 2019, cyberattacks were launched between India and Pakistan, with APT groups targeting each other's classified information, and a separate attack hit India's Kudankulam Nuclear Power Plant, attributed to a North Korea-based group (Khan & Khan, 2025). Advanced Persistent Threats (APTs) now seek to exploit third-party contractors, software updates, and maintenance personnel as a means to penetrate nuclear facilities. Advanced persistent threats (APTs) and insider attacks pose substantial cybersecurity risks to critical systems. APTs involve sophisticated, stealthy attacks by well-funded groups, while insider threats come from employees or contractors with authorized access (Wang, H. et al., 2018). Advanced persistent threats (APTs) pose significant risks to critical infrastructure, including spent fuel management facilities. One known method is by exploiting vulnerabilities in third-party connections, and another protocol is by requiring human adaptability to respond and counter these sophisticated attacks.

A ransomware attack simulation exercise was planned at a nuclear power plant in an eastern European nation. It has experience a surge in cyberattacks amid rising war tensions. Arya, the senior security lead, aimed to test her team's incident response skills (Madhira et al., 2024). However, upon arrival, she discovered a real-life crisis: the control room systems were inaccessible, the cooling system was shut down, and the plant was overheating, risking a nuclear reactor meltdown. The only options were to resolve the issue digitally or perform a manual override, which would deactivate the plant and potentially cause a regional brownout. Nuclear plant personnel remain vulnerable to social engineering, phishing attacks, and zero-day exploits due to a lack of proper cyber wargaming exercises.

Uranium mining facilities face significant cybersecurity challenges due to outdated software, legacy systems, and a lack of robust security measures. Advanced Persistent Threats (APTs) pose significant risks to critical infrastructure, including nuclear facilities, by exploiting vulnerabilities in third-party connections. Cyberattacks on enrichment facilities can have devastating consequences. Wargaming exercises and simulations are crucial to prepare nuclear plant personnel for potential cyber threats, including social engineering, phishing attacks, and zero-day exploits. Implementing encryption, authentication mechanisms, and intrusion detection systems is essential to protect nuclear facilities from cyber assaults.

The Future of Wargaming for Nuclear Security

Nuclear security experts must incorporate wargaming strategies into their cybersecurity frameworks to avoid a digital Chernobyl. This approach will allow nuclear operators to test defenses, simulate attack scenarios, and refine incident response strategies in a safe environment. Experimental wargaming platforms can be utilized to study the impact of tailored nuclear capabilities on international security, finding that lower-yield nuclear weapons may serve as substitutes for higher-yield ones and yield mixed results on conflict escalation (Reddie & Goldblum, 2023). One method for future wargaming for nuclear security is the use of satellite images and wargaming planning to investigate and test scenarios related to nuclear programs collectively. According to Lawrence (2025), western experts and journalists used satellite images to collectively investigate Iran's nuclear program, fostering a shared understanding and persistent questions about its possible military dimensions. Nuclear security experts and cybersecurity personnel can benefit from incorporating red/blue strategies into their cybersecurity frameworks. This framework, strategy, and policy will help to test defenses and improve incident response, while also using tools like satellite images to investigate.

The nuclear industry faced a slowdown following the Fukushima Daiichi accident. However, a renewed interest in zero-carbon electricity and advanced reactors has sparked new development. Digitization in nuclear control rooms involves converting analog signals to digital and creating digital records, enabling modernization. This has created a new framework, called ARUGULA (Boring et al., 2023), which categorizes tasks and performers in nuclear operations, dividing them into human, shared, and automation tasks. This framework considers four stages of information processing by detecting plant status, understanding implications, deciding on a course of action, and acting on that decision. Digitization in nuclear control rooms will help to upgrade analog systems to digital, maintaining existing functionality and human-centered tasks. Online migration, however, shifts functionality to the control system, supporting operators with new features that enhance situation awareness and reduce workload.

The increasingly complex and interconnected nature of nuclear threats demands a robust and collaborative approach. The military is using Enterprise AI to integrate artificial intelligence into its operations. This enhances its intelligence gathering, analysis, and decision-making. Both Ukraine and Russia are leveraging AI to analyze drone and satellite data (Favaro & Williams, 2023). Ukraine's military decided to use facial recognition, natural language processing, and predictive analytics to support its efforts. Emerging technologies introduce new actors, pathways, and timescales into conflict. The increased risk of nuclear use by drawing nuclear reactors into crises may not be avoided. Effective wargaming for nuclear security requires a multilateral cooperation of nuclear agencies, cybersecurity experts, and intelligence services to address the transnational nature of nuclear terrorism.

There must be an agenda to develop comprehensive defense strategies. Thomas Schelling's work on coercion was influenced by the Department of Defense wargames he designed, and his contemporaries also used simulations to study conflict and nuclear scenarios (Lin-Greenberg et al., 2022). This experience and duty allowed him to explore complex strategic interactions. He was able to test hypotheses about conflict dynamics. This early adoption of wargaming and simulation highlights the value of these tools in understanding and navigating complex security challenges.

Nuclear security experts can benefit from incorporating wargaming strategies into their cybersecurity frameworks to test defenses. Wargaming exercises, such as live-fire simulations and scenario-based training, can help nuclear facility personnel prepare for potential cyber threats. Additionally, AI and machine learning models can be utilized to monitor nuclear reactors and detect unusual patterns or unauthorized access attempts. Nonetheless, effective wargaming for nuclear security requires multilateral cooperation, synergy, and mutual support.

Chapter 12: Automotive Cybersecurity

The rapid integration of digital technologies into modern vehicles has significantly expanded the automotive attack surface, introducing new cybersecurity risks. Smart vehicles now rely heavily on interconnected systems that increase their vulnerability to cyber threats and potential exploitation. This chapter analyzes the challenges posed by these expanded attack vectors and the necessity for proactive security measures. Wargaming strategies offer a valuable approach to simulate cyberattacks, assess vulnerabilities, and develop effective defense mechanisms. One or two passages will further explores emerging wargaming methodologies aimed at enhancing the cybersecurity posture of automotive systems in an evolving threat landscape.

Security Automotive

A technological revolution is happening in the automotive industry, with vehicles becoming more connected. In fact, more vehicles are autonomous and dependent on digital infrastructure. The combination of Internet of Things (IoT) and Over-the-Air (OTA) technology in electric vehicles enable benefits such as remote monitoring, software-defined features, and enhanced security (Wang & Khan, 2025). Nowadays, Advanced Driver Assistance Systems (ADAS), over-the-air (OTA) updates, vehicle-to-everything (V2X) communication, and artificial intelligence (AI)-driven automation are standard features in modern cars. In-vehicle communication systems are increasingly vulnerable to cyber threats due to the growing number of sensor-centric devices and computing systems inside vehicles (Rathore et al., 2022). Despite this, existing solutions rely on protocol-specific security techniques and lack a comprehensive security framework.

Hackers could remotely control a compromised vehicle and cause brake failures, unintended acceleration, or even turn off safety systems. Attacks on automotive systems, with millions of smart vehicles on the road, would have the power to disrupt mass traffic, cause financial loss, and even lead to loss of life. Researchers propose developing "smart roads" with sensors to detect collisions, reducing the need for individual vehicle systems and wireless communication except in emergencies, aiming to enhance safety and reduce traffic congestion (Pagale et al., 2025). The potential consequences of cyberattacks on connected vehicles are severe and pose significant safety risks, extending to traffic flow, human life, and infrastructure. Developing smart roads with collision-detection sensors offers a promising solution to mitigate these threats and enhance road safety.

The automotive industry is undergoing a technological revolution with the development of connected and autonomous vehicles. Hackers could remotely control compromised vehicles, causing accidents, financial loss, and loss of life. This highlights the need for comprehensive security frameworks. Researchers propose developing "smart roads" with sensors to detect collisions and enhance safety.

Expanding Attack Surface in Smart Vehicles

As smart vehicles become increasingly connected, they integrate multiple attack vectors. The major entry points for cyber threats are in-vehicle networks, vehicle-to-everything (V2X) communication, autonomous vehicles (AV), and over-the-air (OTA) software Updates.

The increasing connectivity of smart vehicles expands their vulnerability to cyber threats, introducing risks from various entry points. In-vehicle networks, V2X communication, AV systems, and OTA updates each present unique security challenges. These challenges require comprehensive protection measures. As these vehicles rely on complex software, hardware, and connectivity, a single vulnerability in any of these areas could compromise the entire vehicle's security.

In-vehicle network architecture and security are critical to the safe and efficient operation of modern vehicles. The key vehicle components, such as the engine, braking system, and steering, are connected via the controller area network (CAN Bus) (Sharmin et al., 2024). Modern vehicles rely on electronic control units (ECUs) that communicate via the vulnerable controller area network (CAN) protocol, making them susceptible to cyberattacks (Bari et al., 2023). To mitigate this, a machine learning-based intrusion detection system (IDS) using support vector machine (SVM), decision tree (DT), and k-nearest neighbor (KNN) algorithms must be considered. The security and reliability of vehicle-to-everything (V2X) communication systems are critical to the safe and efficient operation of connected and autonomous vehicles. V2X enables the vehicles to communicate with other cars, infrastructure, pedestrians, and cloud servers towards better traffic safety and efficiency. Vehicle-to-everything (V2X) communication devices are used to transmit real-time detection results, perception data, and coordination between vehicles (Chen, H. et al., 2024). Nonetheless, vehicle-to-everything (V2X) communication systems play a vital role in enhancing traffic safety, coordination, and efficiency. These systems assist in real-time data exchange between autonomous vehicles, infrastructure, pedestrians, and cloud servers.

Ultimately, paving the way for smarter and more coordinated transportation systems.

As autonomous vehicles (AVs) become increasingly integrated into public transportation, ensuring safety measures for vulnerable road users (VRUs), such as pedestrians, remains a pressing concern. While existing AV sensor suites excel in vehicle-to-vehicle (V2V) and vehicle-to-infrastructure (V2I) communication, they struggle to identify and track VRUs due to localization difficulties and communication limitations (Yusuf et al., 2024). The current state of V2X and AV technologies is an end-to-end autonomous vehicle motion control architecture that incorporates temporal deep learning algorithms. These algorithms are used to identify and track VRUs more effectively. Other algorithms can evaluate various AI technologies to enhance VRU message sharing, identification, tracking, and communication.

Over-the-Air (OTA) software updates are used to update the firmware on modern vehicles, patching the hardware, upgrading infotainment systems, and applying security fixes. Securing autonomous vehicle software updates is crucial to prevent potential security threats (Qureshi et al., 2022). Over-the-air (OTA) software updates, which are commonly used in autonomous vehicles, are vulnerable to attacks. Black hat hackers will exploit this vulnerability to install malicious software.

The increasing connectivity of smart vehicles expands their vulnerability to cyber threats. Risks arise from various entry points, including in-vehicle networks, vehicle-to-everything (V2X) communication, autonomous vehicles (AVs), and over-the-air (OTA) software updates. In-vehicle networks, relying on the controller area network (CAN) protocol, are susceptible to cyberattacks. This highlights the need for comprehensive protection measures. V2X

113

communication systems play a vital role in enhancing traffic safety and efficiency. Autonomous vehicles pose safety concerns for vulnerable road users (VRUs), and temporal deep learning algorithms. Securing OTA software updates is crucial to prevent potential security threats, as hackers can exploit vulnerabilities to install malicious software.

Future Wargaming Strategies for Automotive Cybersecurity

To keep pace with cyber threats, automakers and security researchers must adopt proactive wargaming strategies that simulate real-world cyberattacks on smart vehicles. These wargames help manufacturers detect vulnerabilities, strengthen their security defenses, and train response teams ahead of a cyber incident. Additionally, these simulations can help identify potential weaknesses in vehicle-to-everything (V2X) communication systems and other connected technologies. Furthermore, wargaming can inform the development of more robust security protocols, procedures, and standards. For the automotive industry, it is strategic to promote a safer and more secure driving experience.

Researchers incorporated artificial intelligence/machine learning (AI/ML) capabilities into a tabletop wargame. According to Tarraf et al., (2025), testing autonomous and remotely operated vehicles in simulated combat scenarios explored potential performance and vulnerabilities of AI/ML-enabled systems. The increasing number of vehicles on the road leads to congestion and accidents (Awais Hassan et al., 2019). These consequences highlight the need for secure vehicle-to-vehicle communication. A decentralized message-passing framework using blockchain technology is proposed, which allows vehicles to communicate securely. Simulations show that the system reduces congestion, increases vehicle speed, and detects malicious vehicles with 77.1% accuracy. For example, a V2X system based on blockchain can prevent attackers from disclosing fake stop signs or green lights, thus preventing accidents. The blockchain-based system provides a secure and distributed network for emergency warning messages and sender authentication.

The automotive industry can enhance smart vehicle security by adopting robust cyber wargaming techniques that simulate real-world cyberattacks. These simulations help identify weaknesses, strengthen security defenses, and inform the development of robust security protocols. Thus, technologies such as blockchain and AI/ML can be integrated into vehicle systems to authenticate messages and detect potential threats, thereby promoting a safer driving experience.

Chapter 13: Wargaming the Future of Aviation Cyber Defense

The aviation industry increasingly relies on complex digital systems, making avionics cybersecurity a critical concern. The expanding cyber threat landscape poses significant risks to the safety, reliability, and operational integrity of aircraft systems. This chapter examines the vulnerabilities inherent in avionics and the challenges these threats present. Wargaming is explored as an effective tool for simulating attack scenarios and evaluating defense strategies in a controlled environment. This section also discusses future wargaming approaches aimed at strengthening aviation cybersecurity and ensuring resilience against emerging threats.

Avionics Cyber Security

Modern aviation is a vital component of national security and global transportation, heavily reliant on digital avionics, automated flight systems, and networked communication technologies. China allegedly reverse-engineered a fighter jet, violating contract terms, and produced its versions (J-11B) with added avionics and weapons (Tyagi et al., 2024). The **RAMSAS** Reliability Analysis by Modeling and Simulation method uses Discrete Event Simulation (DES) for reliability analysis. It's applied in various domains (avionics, automotive, satellites) to model system behaviors and improve reliability (Lazarova-Molnar et al., 2024). Modern aviation plays a critical role in national security and global transportation. These technologies rely heavily on advanced digital technologies. However, intellectual property theft and reverse-engineering pose significant risks. As seen in China's alleged actions with the J-11B fighter jet, involving military-grade reverse engineering techniques, ensuring the reliability and security of aviation systems is vital. Therefore, it is essential to utilize methods like RAMSAS, which can enhance system reliability through simulation-based analysis.

Modern aviation is critical for national security and global transportation, relying on advanced digital technologies such as automated flight systems and networked communications. However, intellectual property theft and reverse-engineering pose significant risks, as seen in China's alleged actions with the J-11B fighter jet. Ensuring the reliability and security of aviation systems is vital to prevent such threats. Methods like RAMSAS (Reliability Analysis by Modeling and Simulation) can help improve system reliability through simulation-based analysis. By using techniques like Discrete Event Simulation (DES), RAMSAS can model system behaviors and enhance reliability in various domains, including flight electronics and satellites.

Wargaming Strategies for Avionics Cybersecurity

To combat the increasing sophistication of cyber threats in aviation, defense organizations, airlines, and aerospace manufacturers must adopt advanced cyber wargaming strategies. One approach is to implement AI-driven anomaly detection for flight systems, enabling real-time monitoring of data by AI-powered cybersecurity systems. Technical measures can enhance unmanned aerial systems (UAS) operations in challenging environments. Additionally, artificial intelligence with deep learning algorithms can facilitate fully autonomous UAS operations and data collection in confined spaces without GPS signals. This measure allows UAS platforms to operate effectively even in direct contact with surveyed structures (Plotnikov & Collura, 2022). Unmanned Aerial Systems (UAS) operations can be enhanced with Quantum cryptography. Quantum cryptography offers theoretically unbreakable encryption, making it an attractive solution for securing UAS communications.

Quantum cryptography for secure flight communications can also be adapted, making it possible to protect cockpit communications, ADS-B transmissions, and satellite uplinks from cyber adversaries. These advancements enhance the security, feasibility, and reliability of quantum communication (Jenefa et al., 2024). For instance, quantum key distribution (QKD) can be utilized to transmit flight control messages unbreakably between pilots and air traffic control (ATC). Quantum communications are beneficial in terms of programmable, latency-aware quantum-secured optical networks, self-correcting quantum control using reinforcement learning, and simulations of quantum key distribution in fiber-based networks.

To counter sophisticated cyber threats in aviation, defense organizations, and aerospace companies should adopt advanced cyber wargaming strategies. Cyber wargames can simulate attacks on aircraft and satellite networks. This enables pilots and cybersecurity teams to develop robust response protocols. Quantum cryptography, including quantum key distribution, can also serve to secure flight communications. Thereby protecting cockpit communications, ADS-B transmissions, and satellite uplinks from cyber adversaries.

Chapter 14: Cyber Threats and the U.S. Space Force

The establishment of the U.S. Space Force marks a pivotal development in national defense, addressing emerging threats in the space domain. As space operations become increasingly critical, the attack surface in space warfare continues to expand, encompassing both cyber and physical vulnerabilities. This chapter explores the unique cybersecurity challenges faced by the Space Force in protecting space assets and infrastructure. Wargaming is presented as a strategic tool to anticipate adversarial tactics, assess risks, and enhance defense capabilities. The discussion further highlights the importance of integrating wargaming into space warfare preparedness to maintain strategic advantage in this evolving domain.

USA Space Force

The chances of cyber warfare, electronic attacks, and kinetic threats aimed at satellites are devastating. Spacecraft and space-based infrastructure are escalating into a different type of warfare between the United States and Russia. In recognition of this reality, the United States established the U.S. Space Force (USSF) as the sixth branch of the military in December 2019. U.S. Space Force (USSF) is tasked with protecting American interests in space through the Department of the Air Force (Galbraith, 2020). The Space Force reflects the growing recognition that space has become a critical domain for military operations. This requires specialized protection and defense strategies. These cybersecurity procedures are designed to safeguard vital satellites and infrastructure from increasingly sophisticated threats.

The Expanding Attack Surface in Space Warfare

NATO member states are increasingly concerned about China's growing space capabilities and potential threats. While the US leads in space power, its dominance within NATO creates vulnerabilities. To address this, NATO is working to increase the role of its allies in space operations. According to Berge and Odgaard (2023), China, a latecomer to the space race, has rapidly developed its capabilities since 2015, creating the People's Liberation.

The army supports strategic orbital force. China is investing in technologies to disrupt space systems. China's military-civil fusion strategy enables collaboration between civilian companies and the military. The US Space Force can look forward to China and its allies developing dual-use capabilities, such as the BeiDou Navigation Satellite System.

A potential long-term merger of cyber and space organizations, including the US Space Force and US Cyber Command, could enhance national defense capabilities. In the short term, efforts are needed to improve cyber requirements and reduce vulnerabilities to outpace adversaries and minimize exploitation risks (Swallow, 2023). So, the Space Force can use wargaming techniques to defend U.S. military satellites, GPS systems, and other critical space assets. While it is more complex, this means the attack surface of cyber and kinetic threats increases. China and Russia have demonstrated advanced anti-satellite capabilities. China has tested ground-to-space systems, including a nuclear-capable hypersonic missile, and created thousands of pieces of orbital debris in a 2007 test (Perron, 2023). Russia has also conducted destructive anti-satellite tests and engaged in suspicious maneuvers. It's been reported that an object was near a US reconnaissance satellite. The object was abnormally maneuvering near US and French-Italian satellites, sparking accusations of espionage and aggressive behavior. These actions highlight the need for international space laws and norms to regulate hacking. Hacking ground control stations is another critical attack strategy. Space-based systems face significant threats from cyberattacks, including satellite hijacking, communication disruption, and redirection into incorrect orbits, as well as hacking of ground control stations.

The Military Operations Research Society (MORS) held a workshop on Campaign Analysis in April 2022, where one working group focused on combining wargaming with modeling and simulation (M&S) to address complex problems (Pournelle, 2024). The synthesis group leader, Trip Barber, emphasized that wargaming is necessary until a comprehensive physics-level model is developed to capture all relevant phenomena of modern combat. Cyber wargaming is essential due to the complexities of cyber operations, which exist in a human-created domain that is not well mapped. Cyber operations are dominated by human choices, involving the manipulation of hardware, software, data, procedures, and people. Wargaming is often the best tool for examining cyber operations, as it allows for the exploration of complex interactions and human decision-making under uncertainty. Computer-based modeling and simulation are best employed for systems in competition with reasonably well-understood phenomena, while wargaming is better suited for addressing complex, human-centered problems. Wargaming is essential for addressing cyber operations and other complex phenomena, such as Space and Electronic Warfare.

The US Space Force was established in 2019 to protect American interests in space from growing threats, including cyberattacks and kinetic threats to satellites. Adversaries such as China and Russia are developing anti-satellite weapons and cyber capabilities, which are increasing the vulnerability of space-based systems. Wargaming is a crucial tool for addressing complex cyber operations and space threats, allowing for the exploration of human decision-making and interactions under uncertainty. Ultimately, the Space Force must use wargaming techniques to defend critical space assets, including satellites and GPS systems.

Chapter 15: Artificial Intelligence and the Future of Wargaming

Artificial intelligence is transforming the landscape of modern conflict by enabling advanced decision-making and autonomous operations. The integration of AI into wargaming offers unprecedented opportunities to simulate complex scenarios with greater accuracy and speed. This chapter examines the role of AI-driven wargaming strategies in anticipating and countering future threats, discussing the challenges associated with implementing AI in military simulations, including ethical considerations and system reliability. This will conclude by exploring how AI will shape the evolution of wargaming and its impact on strategic planning and defense readiness.

Artificial Intelligence

The integration of Artificial Intelligence (AI) is precipitating a paradigmatic shift in modern military strategy. AI influences cybersecurity protocols and tactical decision-making processes, thereby reconfiguring the operational landscape of contemporary warfare. Researchers have discuss the potential of artificial intelligence (AI) in political-military modeling, simulation, and wargaming, particularly in conflicts involving nations with advanced capabilities (Davis & Bracken, 2025). AI will enhance wargaming by providing insights into adversaries' perspectives and decision-making processes. To create a more comprehensive and practical approach to strategic analysis and decision-making, artificial intelligence must be integrated into modeling, simulation, and operational planning to facilitate more informed effective decision-making.

AI Wargaming Strategies for Future Conflicts

As NATO and global adversaries continue to leverage AI-driven military applications, it is imperative that wargaming evolves impactful. This necessitated continued investment and integration by the U.S. Department of Defense (DoD) and other stakeholders. David Jefferson, a young assistant professor, consulted on a RAND Corporation research project for the U.S. Air Force, focusing on speeding up wargaming simulation executions (Fujimoto, 2024). His work contributed to the development of the High-Level Architecture (HLA) for interoperability among US DoD modeling and simulation efforts. The guns, tanks, and missiles won't be the weapons used on the future battlefield, but algorithms, neural networks, and autonomous systems. Defense organizations require AI-powered wargaming strategies to simulate, predict, and preemptively neutralize potential threats. Thus, enhancing proactive defense capabilities.

AI is transforming the complexities of wargaming, revolutionizing how the military approache learning and planning. The artificial intelligence community has utilized real-time strategy games, such as StarCraft II, to develop and test complex algorithms (Goecks et al., 2023). This application has garnered attention from the military research community, as the game, its algorithms, and design explore similar techniques for military scenarios. Various games, including StarCraft II and chess, have been used to develop AI capabilities. The gaming industry's advancements in virtual reality and visual augmentation can potentially enhance battlefield displays for commanders. Researchers aim to bridge the gap between gaming and military applications. Game engines and AI can be leveraged to improve military simulations and decision-making. These technologies will impact future military operations and strategy development.

Media coverage of artificial intelligence (AI) substantially influences public opinion and support for autonomous passenger drones. Researchers have discovered that the relationship between news media attention and trust in AI is mediated by the public's perception of AI explainability (Cheung & Ho, 2025). This indicates that news media significantly influence public understanding and perceptions of AI's explainability. The findings have practical implications. This highlights the importance of news media in educating the public about AI algorithms. The public relies on trust in AI performance when considering support for autonomous passenger drones (Cheung & Ho, 2024). The news media's attention to AI influence society's perceptions of AI explainability, which in turn affects trust in AI. Trust in AI's performance is the key factor driving public support for autonomous passenger drones.

The integration of AI-powered drones, such as those in the Loyal Wingman program, is revolutionizing military air combat by enhancing operational capabilities through coordinated human-machine teaming. For example, Michele Flournoy, former US Under Secretary of Defense, argues that AI will revolutionize warfare (Lushenko & Sparrow, 2024). Defense experts agree that future wars will feature AI-enhanced technologies, including fully autonomous weapons like the US Air Force's "Loyal Wingman" drone, which can identify and engage targets without human oversight.

Traditional battlefield strategies are no longer made obsolete by the use of AI-controlled drone swarms that can overtake an enemy's defenses. AI-powered virtual experiments, like the Loyal Wingman program, introduce realistic unpredictability (Stensrud & Valaker, 2023). This program and similar wargaming technologies necessitate effective human-machine collaboration, founded on shared situational awareness and reciprocal intent prediction.

The integration of artificial intelligence (AI) is transforming modern military strategy, particularly in operational planning, cybersecurity, and tactical decision-making. AI-powered technologies, such as the Loyal Wingman program, are revolutionizing military operations by enhancing human-machine collaboration, hardware, and predictive capabilities. Effective integration of AI into wargaming requires shared situational awareness and mutual intent prediction between humans and machines. As AI continues to shape the future of warfare, defense organizations must invest in AI-driven strategies to stay ahead of potential threats and enhance proactive defense capabilities.

Glossary

Access Control: A security measure that ensures only authorized individuals or systems can access specific resources or data (Dong et al., 2025).

Administrative Controls: Strategies focusing on human factors, awareness, and organizational procedures to manage and mitigate cybersecurity risks (Reuer, 2024).

Advanced Persistent Threats (APTs): A prolonged and targeted cyberattack, often by state-sponsored actors, designed to infiltrate a system and remain undetected (Fei et al., 2025).

Adversarial engagements: Interactions between opposing parties, typically in the context of conflict, where each side seeks to overcome or outmaneuver the other (Holburn et al., 2024).

Adversarial Simulation: Role-playing exercises that simulate the actions of competitors or hostile actors to understand better their potential moves and reactions (Oh et al., 2023).

AI-driven simulations: Simulations powered by artificial intelligence that predict and replicate scenarios to improve decision-making (Biriukov & Vácha, 2024).

American Revolution: The war fought between the Thirteen American Colonies and Great Britain from 1775 to 1783, resulting in American independence (Papworth & Dence, 2024).

Antivirus Software: A program that scans and removes malware from systems, ensuring the security of data and devices against malicious attacks (Belal & Sundaram, 2024).

Audit and Verification: The process of checking records, filtering suspicious accounts, and verifying data integrity (Anderson et al., 2024).

Audit Logging: The process of recording and analyzing actions taken on a system to detect security breaches or unauthorized activity (Bates et al., 2015).

Audit Purposes: The use of wargaming kits or simulations enables the recording and review of actions and decisions made during an exercise, facilitating later analysis and lessons learned (Yoon et al., 2024).

Automatic Dependent Surveillance Broadcast (ADS-B): ADS-B broadcasts aircraft location data without encryption, making it susceptible to spoofing and cyber manipulation (Radmanesh et al., 2016).

Automotive Ethernet: A high-speed network standard that is becoming more prevalent in vehicles but also introduces new security vulnerabilities (Douss et al., 2023).

Backup Strategy: A plan for regularly backing up data to ensure its availability and integrity in case of system failure or a cyberattack (Yu et al., 2024).

Behavioral Feedback Loop: The process by which the outcomes of actions in a wargame influence future behavior, enabling participants to adjust their strategies or responses based on the feedback received (Mergel, 2024).

Behavioral Models: Analytical frameworks that focus on the behaviors and motivations of participants (e.g., competitors, adversaries) in a wargame, rather than just technical or algorithmic solutions, to predict actions and outcomes (Wang, A. Z. et al., 2025).

Black Swans: Unpredictable and high-impact events that have significant consequences, often explored in wargaming scenarios (Hoey, 2025).

Blue teaming: The defensive counterpart to red teaming, where a group of security professionals defends against attacks, simulating the response to real-world cyber threats (Li, L. et al., 2025).

Bug bounty programs: Cybersecurity initiatives that reward individuals for discovering and reporting security vulnerabilities in a system (Li, X. et al., 2023).

Business Wargaming: A structured, strategic exercise designed to simulate competitive or adversarial scenarios in business environments (Spaniol, 2024).

Chain of Custody: The documentation and process that ensures evidence collected during an investigation is handled correctly and preserved for legal purposes (Tan, 2024).

Clearance Process: The procedure for granting access to information or systems based on an individual's clearance level (Addo, 2022).

Code Injection: A method used by attackers to insert malicious code into a program, often exploiting a vulnerability, to manipulate its behavior or extract information (Kakisim, 2024).

Cognitive Modeling: The process of creating a model of human thought and decision-making processes (Lee, M.D., 2024).

Compensating Controls: Alternative measures used when primary controls are not feasible or effective (Slayton, 2021).

Competitors: Other organizations in the same industry or market that influence business practices, strategies, and security measures (Nayak et al., 2022).

Compliance and Certification: The process of training, certifying, and updating administrators on relevant cybersecurity frameworks to ensure adherence to security standards (Moses et al., 2025).

Compliance Framework: A set of guidelines and standards that organizations follow to ensure adherence to legal, regulatory, and industry-specific cybersecurity requirements (Amal Chandra, 2024).

Controller Area Network (CAN Bus): A vehicle communication protocol used to link Electronic Control Units (ECUs) in modern vehicles, making them vulnerable to cyberattacks (Sharmin et al., 2024).

Corrective Controls: Actions taken after a security incident to mitigate damage and restore normal business operations (Jule, 2020).

Critical infrastructure protection: Efforts to secure essential services and systems, such as energy, water, transportation, and communications, from cyber threats and attacks (Katkuri, 2024).

Crowdsourcing: The practice of obtaining input, ideas, or solutions from a large group of people, often through an open-call model (Moghadasi et al., 2024).

Cryptography: The practice and study of techniques for securing communication and information through codes, ensuring confidentiality and integrity (Aldosari & Aldawsari, 2024).

Cyber capabilities: The technological and strategic tools, practices, and resources used to defend or attack in the cyber domain (Arsenault et al., 2024).

Cyber intrusion: Unauthorized access to or disruption of a computer system or network by malicious actors (Tyagi et al., 2024).

Cyber Operations: Military or strategic operations involving the use of cyberspace to target and disrupt systems, typically in the context of cyber warfare (Segate, 2024).

Cyber Warfare: The use of digital attacks by one nation-state against another to cause disruption, espionage, or sabotage (Jung, 2024).

Cyber Wargaming Framework: A strategic approach that combines traditional wargaming principles with a focus on cybersecurity, designed to test an organization's resilience against cyber threats (Colbert et al., 2020).

Cyber physical security: The protection of critical infrastructure, such as power grids, transportation systems, and communication networks, which relies on both digital and physical security measures (Kanovich et al., 2017).

Cybersecurity Ecosystem: The network of tools, software, policies, and practices to protect data and systems (Amal Chandra, 2024).

Data breaches: The unauthorized use of viruses, malware, and data destabilization can wreak havoc or manipulate official statements, messages, or networks (Chua, 2021).

Data Protection: Safeguarding data from modification, disruption, destruction, or unauthorized disclosure (Sodhi et al., 2024).

Data Scrambling: A technique used to make sensitive data unreadable to unauthorized parties by transforming it into a scrambled format, often used to protect data during storage or transmission (Xu, Z. et al., 2023).

DDoS (Distributed Denial-of-Service) Attack: A cyberattack aimed at overwhelming a system or network by flooding it with excessive traffic, causing it to crash or become unavailable (Muthukumar & Ashfauk Ahamed, 2024).

Debugging Tools: Software utilities that aid in identifying and resolving bugs or issues in systems and software, enabling administrators to pinpoint errors and malfunctions (Bellucci et al., 2024).

Decision Tree (DT): A machine learning algorithm used for decision-making processes, applicable in intrusion detection (Benítez-Andrades et al., 2024).

Deep Q-Learning: A type of reinforcement learning that utilizes deep neural networks to solve complex decision-making problems in environments with large state spaces, often applied in cybersecurity scenarios, such as protecting power grids (Cheng et al., 2024).

Defense contractor. A defense contractor is any entity, individual, firm, corporation, partnership, association, or other legal non-federal entity that enters a contract directly with the Department of Defense to furnish services, supplies, or construction supported by 32 Code of Federal Regulations 158.3 definitions (Li, Y, 2018).

Detective Controls: Tools or strategies used to detect and identify irregularities or breaches after they occur (Grimaila et al., 2012).

Deterrent Controls: Security measures designed to discourage or prevent potential attackers from attempting to breach an organization's systems or policies (Leeuw & Leeuw, 2012).

Digital Twin: A digital twin is a real-time virtual model of a physical system used for simulation and performance optimization (Al-Qirim et al., 2025).

Digital-Twin Simulation: A digital representation of a physical system or environment used for simulation, testing, and real-time data analysis, especially in robotic systems (AbdElSalam et al., 2025).

Disaster Recovery Plans: Strategies designed to recover and restore data, systems, and networks after a cybersecurity incident or physical disaster, minimizing downtime and data loss (Maggu et al., 2025).

Dynamic Simulation: A method used to predict and visualize the impact of various cybersecurity events, enabling administrators to anticipate how a data breach might unfold and prepare effective responses accordingly (Jain et al., 2024).

Electronic Control Units (ECUs): Computers within a vehicle that control various vehicle components, such as the engine, brakes, and airbags, and can be vulnerable to cyberattacks (Esenduran et al., 2025).

Enterprise Standards: Company-wide policies and standards for using resources and technologies securely (Rühlig & Ten Brink, 2021).

Ethical hacking: The practice of legally probing a system for security flaws to help identify and fix vulnerabilities before they can be exploited by malicious hackers (Selvarajan et al., 2025).

Evasion Techniques: Strategies employed by attackers to bypass security defenses or detection systems, making it more difficult to identify or mitigate their malicious actions (Román et al., 2025).

External Influences: Factors originating outside an organization that impact its operations, controls, and security strategies (Tribou et al., 2025).

False Negative Rate: A measure of how often a system fails to detect an actual positive instance, relevant for intrusion detection systems (Luo et al., 2024).

Feedback Systems: Mechanisms in technical controls that validate and route data back into the system for processing (Wei et al., 2025).

Firewalls: Network security devices or software designed to monitor and control incoming and outgoing network traffic based on predetermined security rules, acting as a barrier between trusted and untrusted networks (Boopathi, 2025).

Fly-By-Wire (FBW): FBW replaces mechanical flight controls with electronic systems, increasing automation but introducing cybersecurity vulnerabilities (Skraaning Jr & Jamieson, 2024).

Game Theory: A mathematical framework used to model strategic interactions between different decision-makers (players), often applied in wargaming to simulate competitive or adversarial scenarios (Reid et al., 2025).

Incident response: The process of identifying, managing, and mitigating the impact of security incidents or breaches, such as cyberattacks (Fu et al., 2024).

Incident Response Plan: A documented strategy outlining the steps to take during and after a security incident to minimize damage and restore operations (Datta & Acton, 2024).

Incident Response Playbook: A predefined set of actions and procedures that an organization follows to manage and mitigate the effects of a security incident, such as a data breach (Madhira et al., 2024).

Inclusiveness: The practice of ensuring diverse participation and access to resources within a strategy or practice (Xu, Y. et al., 2025).

Industrial Control Systems (ICS): Systems used to control industrial processes such as manufacturing, power generation, and water treatment. These systems are vulnerable to cyber threats (Al-Qirim et al., 2025).

Ineffectiveness Measure: Metrics used to assess whether specific wargaming strategies, tools, or equipment are not yielding the desired outcomes, allowing for adjustments or reevaluations of the approach (Lee, J. 2025).

Information Security Regulations: Legal and organizational rules and guidelines designed to protect data and systems from unauthorized access, modification, or destruction (Al-Khatib et al., 2025).

Innovative Thinking: The process of brainstorming new solutions or techniques (Qi, 2024).

Internal Culture: The shared values, beliefs, and practices within an organization that affect its operations and decision-making (Deja, 2024).

Internal Influences: Factors originating within an organization that impact its operations, controls, and security strategies (Xu, J., 2024).

Intrusion Detection Mechanism: Systems or techniques designed to identify unauthorized access or anomalies in a computer network or infrastructure (Prasath et al., 2024).

Intrusion Detection System (IDS): A system designed to detect unauthorized access or anomalies within a network, such as the in-vehicle network of a vehicle (Al-Qirim et al., 2025).

Kali Linux: A Debian-based Linux distribution used for penetration testing, ethical hacking, and network security assessments, providing various tools for these purposes (Danesh et al., 2024).

King Philip's War: A conflict that occurred between 1675 and 1678 in New England, primarily between Native American tribes led by Metacom (known as King Philip) and English colonists. The war was marked by violent confrontations and was one of the bloodiest conflicts in early American history (Trigg & Mrozowski, 2024).

K-Nearest Neighbor (KNN): A machine learning algorithm used for classification by analyzing the closest data points, used in intrusion detection systems (Liu, X. et al., 2024).

Live-fire exercises: Realistic training simulations where participants engage in live-action scenarios (Mensen et al., 2024).

Malware: Software designed to harm, exploit, or disrupt a computer system or network (TM, & Taurshia, n.d.).

Military strategy: The planning and execution of military operations to achieve specific goals, often involving tactics, resources, and positioning to outmaneuver an adversary (Lee, O. et al., 2024).

Mitigation Strategies: Actions or countermeasures implemented to reduce or prevent the impact of potential cyberattacks or threats (Nieuwenhuizen, 2024).

Multi Factor Authentication (MFA): A security system requiring multiple forms of verification before granting access to a system (Shukla & Patel, 2022).

Need-to-Know Access: A security principle that grants access to sensitive information only to individuals with specific clearance (López Velásquez et al., 2023).

Need-to-Know Basis: A principle that restricts access to information to individuals who need it to perform their specific job functions (Mensah et al., 2024).

Network Connectivity Issues: Problems that arise when communication between devices is disrupted, often requiring specialized debugging tools to resolve (Farooq et al., 2024).

NIST Guidelines: Standards and guidelines from the National Institute of Standards and Technology for effective cybersecurity (Sattar et al., 2021).

Openness Strategy: A communication approach that encourages transparency and free exchange of information between internal teams and external stakeholders (Huang et al., 2023).

Over-the-Air (OTA) Updates: The wireless transmission of software updates to devices, particularly in vehicles, can introduce vulnerabilities if not properly secured (Gu et al., 2024).

Palo Alto Firewalls: A specific brand of firewall systems known for providing advanced threat prevention and network security, often used in enterprise environments to protect against various types of cyberattacks (Toga et al., 2024).

Patch Management: The process of managing, applying, and verifying patches or updates to software and systems to address vulnerabilities (Radanliev et al., 2024).

Penetration testing: A method of testing a computer system, network, or web application to identify security vulnerabilities by simulating an attack from malicious outsiders (Petty et al., 2024).

Personal Identifiable Information (PII): Any data that can be used to identify a person, such as their name, address, or social security number (Leong et al., 2024).

Physical Controls: Security measures that physically restrict access to systems, data, or buildings (Lo, 2024).

Preventative Controls: Security measures implemented to prevent security breaches or incidents from occurring in the first place (Parkin & Chua, 2022).

Quantum Key Distribution (QKD): QKD uses quantum mechanics to securely transmit encryption keys, detecting interception attempts automatically (Jayashree, Preprint).

Ransomware Attack: A type of cyberattack where attackers encrypt the victim's data and demand payment for decryption keys (Li, X. et al., 2024).

Red teaming: A cybersecurity approach where a group of security professionals simulates an attack on a system, organization, or network (Frank, 2025).

Regulatory Controls: Mandatory legal and compliance requirements that organizations must adhere to in order to meet standards and regulations (Currie & Seddon, 2024).

Reverse Engineering: The process of deconstructing software or hardware to understand its design, operation, and vulnerabilities, often employed by attackers to identify potential weaknesses (Adomako et al., 2024).

Risk Management Framework (RMF): A structured approach to managing and mitigating risks, particularly in cybersecurity (Mentzas et al., 2024).

SaaS Tools (e.g., DataDog): Software-as-a-Service (SaaS) platforms used for monitoring and analyzing the performance of systems, networks, and applications in real-time, offering insights into security, reliability, and efficiency (Hannay et al., 2021).

SCARA Robots: Selective Compliance Assembly Robot Arm, a type of industrial robot known for its efficiency in assembly and manufacturing processes (Zhen et al., 2023).

Scenario Planning: A method for predicting and planning for future uncertainties in business or cybersecurity, focusing on potential outcomes and strategic decision-making in response to risks, including cyber threats (Sohrabinejad et al., 2025).

Security Awareness Training: Educational programs designed to enhance employees' understanding of cybersecurity risks and best practices for mitigating those risks (Katsarakes et al., 2024).

Security Clearance: The level of access granted to an individual based on their role and certification (Sarkar, S. et al., 2024).

Security Culture: The shared attitudes, practices, and policies within an organization that influence employees' perceptions and actions regarding security (Storbeck et al., 2025).

Security Group: A set of permissions and roles that define access levels and control over resources within a system or network (Abu Al et al., 2022).

Security posture: The overall security status of an organization, which includes its ability to protect against cyber threats and respond to potential attacks (Jayashree, Preprint).

Shapley Additive Explanations (SHAP): SHAP explains the contribution of each input feature in machine learning models, improving transparency and trust in cybersecurity (Gurmessa & Jimma, 2025).

Sole contracts: Procurement contracts awarded to a single supplier, often used in government or military acquisitions, sometimes due to the urgency or specificity of the need (Iossa et al., 2022).

Stakeholders: Individuals or groups who have an interest in the outcomes of a wargame, such as competitors, customers, suppliers, regulators, and investors, all of whom influence or are affected by the game's results (Uysal & Sandıkkaya, 2025).

State-sponsored cyber activities: Cyberattacks or cyber espionage conducted by government-affiliated groups or state actors, typically for political, military, or economic gain (Perdana et al., 2024).

Stochastic Game: A game-theoretic model used to represent scenarios with probabilistic elements, particularly useful for decision-making in uncertain environments (Chen & Kan, 2024).

Strategic Alignment: The process of ensuring that all actions, decisions, and efforts within an organization are consistent with its overall goals and objectives, particularly in cybersecurity strategy and wargaming (Fantozzi et al., 2025).

Strategic cyberattacks: Deliberate cyber operations designed to achieve strategic goals, often at a national or geopolitical level, such as espionage or sabotage (Shao & Li, 2022).

Strategy Process: A decision-making framework used to secure assets, developed through collaboration and trust, and guided by verified data. It includes high-level planning and may evolve through principles or specific implementation strategies (Amdam & Benito, 2022).

Stuxnet: A highly sophisticated malware designed to target industrial control systems, particularly those used in nuclear facilities, and cause physical damage (Bruschi et al., 2024).

Supervisory Control and Data Acquisition (SCADA): A type of industrial control system used for monitoring and controlling industrial processes, which can be vulnerable to cyberattacks (Giustozzi et al., 2024).

Support Vector Machine (SVM): A machine learning algorithm used for classification tasks, such as intrusion detection in vehicle networks (Ye et al., 2024).

System Restore: A feature in many systems that allows users to revert their computer's state to a previous point in time, typically used to recover from system failures or malware infections (Cupek et al., 2017).

Technical Controls: Security solutions involving hardware or software to protect against cyber-attacks (Albayrak & Bağcı, 2025).

Theoretical Assessments: Evaluations conducted by red and blue teams that involve analyzing and responding to hypothetical cyber threats, used to refine strategies and prepare participants for real-world situations (Skowronek et al., 2025).

Three Warfares strategy: A Chinese strategy involving the use of psychological, legal, and media operations to undermine adversaries, including cyberattacks targeting military, economic, and diplomatic sectors (Bisht et al., 2019).

Transparency: In cybersecurity, transparency refers to clear and open communication, including coordinated vulnerability disclosures, that allows for proper assessment of a system's security status and fosters accountability (Kurteva et al., 2024).

True Positive Rate: A measure of how well a system detects actual positive instances, relevant for evaluating intrusion detection systems (Sethuraman et al., 2024).

Unified Combatant Command: A joint military command structure in the U.S. Department of Defense that integrates all military branches for coordinated action, including cyber operations (Perron, 2023).

Vehicle-to-Everything (V2X) Communication: V2X enables vehicles to communicate with infrastructure, pedestrians, and other vehicles to enhance road safety and traffic efficiency (Liu, H. et al., 2024).

Vulnerability Monitoring: The continuous process of scanning and identifying security vulnerabilities in systems and networks (Bai et al., 2022).

Vulnerability Testing: The process of identifying weaknesses in a system, network, or application that could be exploited by malicious actors (Karthika & Rangasamy, 2024).

Wargaming: A structured, scenario-based simulation designed to model conflicts, attacks, and defensive strategies, enhancing resilience, evaluating decision-making, and improving cybersecurity preparedness (Wong et al., 2025).

Wargaming Scenarios: Specific exercises or simulations designed to replicate real-world cyber threats and test responses to potential attacks, focusing on various situations that may arise during a cyber incident (Ryseff & Bond, 2025).

Web Vulnerability Scanning Tools: Automated software tools used to identify security weaknesses in web applications and websites, such as SQL injection, cross-site scripting (XSS), and other vulnerabilities (Román et al., 2025).

Zero-day exploits: Cyberattacks that target vulnerabilities in software or systems that are unknown to the vendor or have not been patched, leaving systems exposed (Akpaku et al., 2025).

Zero-day Vulnerabilities: Security flaws in software or systems that are unknown to the vendor and have no available patch, making them highly exploitable by attackers (Datta & Acton, 2024).

Zero-Sum Nash Strategy: A strategy from game theory used to describe a situation in which one player's gain or loss is exactly offset by the losses or gains of another player, often applied in cybersecurity to model defense strategies (Guevara, 2025).

References

AbdElSalam, M., Bensalem, S., Delacourt, A., He, W., Katsaros, P., Kekatos, N., Nolasco Ruiz, R., Peled, D., Ponchant, M., Ryad, I., Temperekidis, A., & Wu, C. (2025). Digital twin for the formal analysis of a depth of anesthesia controller. *SIMULATION, 101* (3), 341-360. https://doi.org/10.1177/00375497241311617

Abu Al, H. Q., Al Badawi, A., & Bojja, G. R. (2022). Boost-Defence for resilient IoT networks: A head-to-toe approach. Expert Systems, 39(10), 1–15. https://doi.org/10.1111/exsy.12934

Addo, A. (2022). Information technology and public administration modernization in a developing country: Pursuing paperless clearance at Ghana customs. *Information systems journal, 32* (4), 819-855.

Adomako, S., Gyensare, M. A., Amankwah, A. J., Akhtar, P., & Hussain, N. (2024). Tackling grand societal challenges: Understanding when and how reverse engineering fosters frugal product innovation in an emerging market. *Journal of Product Innovation Management, 41* (2), 211–235. https://doi.org/10.1111/jpim.12678

Ahmad, M. O., Tripathi, G., Siddiqui, F., Alam, M. A., Ahad, M. A., Akhtar, M. M., & Casalino, G. (2023). BAuth-ZKP-A blockchain-based multi-factor authentication mechanism for securing smart cities. *Sensors (Basel, Switzerland), 23* (5), 2757. https://doi.org/10.3390/s23052757

Ahmed, M., & Gaber, M. (2024). An investigation on the cyber espionage ecosystem. *Journal of Cyber Security Technology*, 1–25.

Albayrak, E., & Bağcı, H. (2025). Modelling the effects of personal factors on information security awareness. *Journal of Information Science, 51* (1), 20-30.

Aldosari, S. S., & Aldawsari, L. S. (2024). PQ-LEACH: A novel post-quantum protocol for securing WSNs communication. *International Journal of Engineering Business Management, 16*, 18479790241301163.

Al-Janabi, S., Jabbar, H., & Syms, F. (2024). Cybersecurity transformation: Cyber-resilient IT project management framework. *Digital, 4* (4), 866–897. https://doi.org/10.3390/digital404004

Al-Khatib, S. F., Ibrahim, Y. Y., & Alnadi, M. (2025). Cybersecurity Practices and Supply Chain Performance: The Case of Jordanian Banks. *Administrative Sciences (2076-3387), 15* (1), 1. https://doi.org/10.3390/admsci15010001

Al-Qirim, N., Majdalawieh, M., Bani-hani, A., & Al Hamadi, H. (2025). Cyber threat intelligence for smart grids using knowledge graphs, digital twins, and hybrid machine learning in SCADA networks. *International Journal of Engineering Business Management, 17*, 18479790251328183

Akpaku, E., Chen, J., Ahmed, M., Ayitey Sosu, R. N., Agbenyegah, F. K., & Kofi Louis, D. (2025). eBiTCN: Efficient bidirectional temporal convolution network for encrypted malicious network traffic detection. *Journal of Computer Security*, 0926227X251326282.

Amal Chandra, C. (2024). Strengthening India's Cybersecurity and Data Privacy Landscape: A Comprehensive Overview. *Indian Journal of Public Administration, 70* (3), 466-478.

Amdam, R. P., & Benito, G. R. G. (2022). Opening the Black Box of International Strategy Formation: How Harvard Business School Became a Multinational Enterprise. *Academy of Management Learning & Education, 21*(2), 167–187. https://doi.org/10.5465/amle.2020.0028

Anderson, S. J., K. Chintagunta, P., & Vilcassim, N. (2024). Virtual Collaboration Technology and International Business Coaching: Examining the Impact on Marketing Strategies and Sales. *Marketing Science, 43* (3), 637–672. https://doi.org/10.1287/mksc.2019.0121

Arsenault, A. C., Kreps, S. E., Snider, K. L., & Canetti, D. (2024). Cyber scares and prophylactic policies: Crossnational evidence on the effect of cyberattacks on public support for surveillance. *Journal of Peace Research, 61* (3), 413-428.

Ashraf, M. (2022). The role of peer events in corporate governance: Evidence from data breaches. *The Accounting Review, 97* (2), 1-24. https://doi.org/10.2308/TAR-2019-1033

Azubuike, C. F. (2023). Cyber security and international conflicts: An analysis of state-sponsored cyber-attacks. *Nnamdi Azikiwe Journal of Political Science*, 8(3), 101–114.

Awais Hassan, M., Habiba, U., Ghani, U., & Shoaib, M. (2019). A secure message-passing framework for inter-vehicular communication using blockchain. *International Journal of Distributed Sensor Networks, 15*(2), 1550147719829677.

Bai, J., Zhang, Z., & Shen, B. (2022). Internet of vehicles security situation awareness based on intrusion detection protection systems. *Journal of Computational Methods in Science and Engineering, 22*(1), 189-195.

Barati, M., & Yankson, B. (2022). Predicting the occurrence of a data breach. *International Journal of Information Management Data Insights, 2*(2), 100128. https://doi.org/10.1016/j.jjimei.2022.100128

Bari, B. S., Yelamarthi, K., & Ghafoor, S. (2023). Intrusion detection in vehicle controller area network (CAN) bus using machine learning: A comparative performance study. *Sensors (Basel, Switzerland), 23*(7), 3610. https://doi.org/10.3390/s23073610

Bates, A., Butler, K. R., Sherr, M., Shields, C., Traynor, P., & Wallach, D. (2015). Accountable wiretapping–or–I know they can hear you now. *Journal of Computer Security, 23*(2), 167-195.

Baumann, L., & Utz, S. (2021). Professional networking: Exploring differences between offline and online networking. *Cyberpsychology, 15*(1) doi: 10.5817/CP2021-1-2

Belal, M. M., & Sundaram, D. M. (2024). Multi-variants vision transformer-based malware image classification model using multi-criteria decision-making. *Journal of Intelligent & Fuzzy Systems, 46*(5-6), 11331-11351.

Bellucci, M., Delestre, N., Malandain, N., & Zanni-Merk, C. (2024). Towards counterfactual explanations for ontologies. *Semantic Web, 15*(5), 1611-1636.

Benaroch, M. (2019). IT service providers and cybersecurity risk. *The Armed Forces Comptroller, 64*(4), 50-54.

Benítez-Andrades, J. A., Prada-García, C., García-Fernández, R., Ballesteros-Pomar, M. D., González-Alonso, M. I., & Serrano-García, A. (2024). Application of machine learning algorithms in classifying postoperative success in metabolic bariatric surgery: A comprehensive study. *Digital Health, 10*, 20552076241239274.

Berge, J. V., & Odgaard, L. (2023). NATO in the Global Commons: Defending Outer Space Against Threats from China. *International Journal, 78*(4), 634-642.

Blaschke, L. M. (2021). The dynamic mix of heutagogy and technology: Preparing learners for lifelong learning. *British Journal of Educational Technology, 52*(4), 1629–1645.

Biriukov, D., & Vácha, R. (2024). Pathways to a shiny future: Building the foundation for computational physical chemistry and biophysics in 2050. *ACS Physical Chemistry Au, 4* (4), 302-313. https://doi.org/10.1021/acsphyschemau.4c00003

Bisht, N. S., Jain, R., & Gambhir, V. (2019). Doklam plateau and three warfares strategy. *China Report, 55*(4), 293-309.

Boopathi, M. (2025, February). Hybrid optimization based deep stacked autoencoder for routing and intrusion detection. In *Web Intelligence* (Vol. 23, No. 1, pp. 3-22). Sage UK: London, England: SAGE Publications.

Boring, R. L., Ulrich, T. A., & Lew, R. (2023, September). Levels of digitization, digitalization, and automation for advanced reactors. In *Proceedings of the Human Factors and Ergonomics Society Annual Meeting* (Vol. 67, No. 1, pp. 207-213). Sage CA: Los Angeles, CA: SAGE Publications.

Bossone, A. (2018). The battle against breaches: A call for
modernizing federal consumer data security regulation.
Federal Communications Law Journal, 69(3), 227-199.

Bruschi, D., Di Pasquale, A., Lanzi, A., & Pagani, E. (2024).
Ensuring cybersecurity for industrial networks: A solution for
ARP-based MITM attacks. *Journal of Computer Security, 32*
(5), 447-475.

Brynen, R. (2020). Virtual paradox: How digital war has reinvigorated
analogue wargaming. *Digital War, 1*(1-3), 138-143. doi
:10.1057/s42984-020-00004-z

Büchler, J. P. (2022). *Business Wargaming for Mergers
&mAcquisitions:Systematic Application in the Strategy and
Acquisition Process.* Springer Nature.

Budning, K., Wilner, A., & Cote, G. (2022). From physical to virtual
to digital: The Synthetic Environment and its impact on
Canadian defence policy. *International Journal, 77*(2), 335-
355

Burt, S. K. (2023). President Obama and China: Cyber diplomacy
and strategy for a new era. *Journal of Cyber Policy,* 8(1), 48–
66.

Caron, F. (2021). Obtaining reasonable assurance on cyber resilience.
Managerial Auditing Journal, 36(2), 193-217.
https://doi.org/10.1108/MAJ-11-2017-1690

Cassottana, B., Roomi, M. M., Mashima, D., & Sansavini, G. (2023).
Resilience analysis of cyber-physical systems: A review of
models and methods. Risk Analysis: An International Journal,
43(11), 2359–2379. https://doi.org/10.1111/risa.14089

Chaskes, W. (2022). The Three Laws: The Chinese Communist Party Throws Down the Data Regulation Gauntlet. *Wash. & Lee L. Rev.*, 79, 1169.

Chen, H., Bandaru, V. K., Wang, Y., Romero, M. A., Tarko, A., & Feng, Y. (2024). Cooperative perception system for aiding connected and automated vehicle navigation and improving safety. *Transportation Research Record, 2678*(12), 1498-1510.

Chen, J., Henry, E., & Jiang, X. (2022). Is cybersecurity risk factor disclosure informative? Evidence from disclosures following a data breach. *Journal of Business Ethics*, https://doi.org/10.1007/s10551-022-05107-z

Chen, Z., & Kan, Z. (2024). Real-time reactive task allocation and planning of large heterogeneous multi-robot systems with temporal logic specifications. *The International Journal of Robotics Research*, 02783649241278372.

Cheng, C., Li, G., & Fan, J. (2024). Deep Q learning cloud task scheduling algorithm based on improved exploration strategy. *Journal of Computational Methods in Science and Engineering, 24* (4-5), 2095-2107.

Cheung, J. C., & Ho, S. S. (2024). Explainable AI and trust: How news media shapes public support for AI-powered autonomous passenger drones. *Public Understanding of Science*, 09636625241291192.

Cheung, J. C., & Ho, S. S. (2025). Explainable AI and trust: How news media shapes public support for AI-powered autonomous passenger drones. *Public Understanding of*

Science (Bristol, England), 34 (3), 344–362.
https://doi.org/10.1177/09636625241291192

Chua, J. A. (2021). Cybersecurity in the healthcare industry. *The Journal of Medical Practice Management, 36*(4), 229-231.

Clapper, J., Lettre, M., & Rogers, M. S. (2017). Foreign cyber threats to the United States. *Hampton Roads International Security Quarterly,* 1.

Coenraad, M., Pellicone, A., Ketelhut, D. J., Cukier, M., Plane, J., & Weintrop, D. (2020). Experiencing cybersecurity one game at a time: A systematic review of cybersecurity digital games. *Simulation & Gaming, 51* (5), 586-611. doi :10.1177/1046878120933312

Colbert, E. J., Kott, A., & Knachel, L. P. (2020). The game-theoretic model and experimental investigation of cyber wargaming. *Journal of Defense Modeling and Simulation, 17* (1), 21-38. https://doi.org/10.1177/1548512918795061

Currie, W. L., & Seddon, J. J. (2024). Competing stakeholder narratives on crypto-assets: Miracle or mirage?. *Journal of Information Technology, 39* (2), 339-360.

Cupek, R., Folkert, K., Fojcik, M., Klopot, T., & Polakow, G. (2017). Performance evaluation of redundant OPC UA architecture for process control. *Transactions of the Institute of Measurement and Control, 39* (3), 334-343.

Danesh, H., Karimi, M. B., & Arasteh, B. (2024). CMShark: A NetFlow and machine-learning based crypto-jacking intrusion-detection method. *Intelligent Decision Technologies, 18* (3), 2255-2273.

Datta, P. M., & Acton, T. (2024). Did a USB drive disrupt a nuclear program? A Defense in Depth (DiD) teaching case. *Journal of Information Technology Teaching Cases, 14* (2), 311-321.

Davis, P. K., & Bracken, P. (2025). Artificial intelligence for wargaming and modeling. *Journal of Defense Modeling and Simulation, 22* (1), 25–40. https://doi.org/10.1177/15485129211073126

Deja, M. (2024). Information culture of university administration: Making personnel bureaucracy a professional bureaucracy. *Journal of Librarianship and Information Science, 56* (2), 379-396.

Dong, J., Jiang, R., Zhang, Y., Tian, S., & Wu, B. (2025). A hierarchical access control and sharing model for healthcare data with on-chain-off-chain collaboration. *Journal of Intelligent & Fuzzy Systems, 48* (3), 215-231.

Douss, A. B. C., Abassi, R., & Sauveron, D. (2023). State-of-the-art survey of in-vehicle protocols and automotive ethernet security and vulnerabilities. *Mathematical Biosciences and Engineering : MBE, 20* (9), 17057-17095. https://doi.org/10.3934/mbe.2023761

Efe, A. (2023). A comparison of key risk management frameworks: COSO-ERM, NIST RMF, ISO 31.000, COBIT. *Denetim Ve Güvence Hizmetleri Dergisi*, 3(2), 185–205.

Efrony, D., & Shany, Y. (2018). A rule book on the shelf? Tallinn manual 2.0 on cyberoperations and subsequent state practice. *The American Journal of International Law, 112* (4), 583–657. https://doi.org/10.1017/ajil.2018.86

Esenduran, G., Jin, M., & Zhou, Y. (2025). Laissez-Faire vs. Government Intervention: Implications of Regulation Preventing Nonauthorized Remanufacturing. Manufacturing & Service Operations Management (M&SOM), 27(2), 588–606. https://doi.org/10.1287/msom.2023.0128

Ettinger, A. (2018). Trump's national security strategy: "America first" meets the establishment. *International Journal (Toronto), 73*(3), 474–483. https://doi.org/10.1177/0020702018790274

Fagiolini, A., Dini, G., Massa, F., Pallottino, L., & Bicchi, A. (2024). Distributed misbehavior monitors for socially organized autonomous systems. *The International Journal of Robotics Research, 43*(14), 2145-2182.

Fantozzi, I. C., Olhager, J., Johnsson, C., & Schiraldi, M. M. (2025). Guiding organizations in the digital era: Tools and metrics for success. *International Journal of Engineering Business Management, 17,* 18479790241312804.

Farooq, W., Baig, N., Khan, B. A., Butt, F. A., Hanif, A., Ali, A., & Raza, M. R. (2024). Enhancement of paediatric oncology pharmacy practices in a low–middle-income country through teaching and training using the My Child Matters Grant. *Journal of Oncology Pharmacy Practice, 30*(5), 786-791.

Favaro, M., & Williams, H. (2023). False sense of supremacy: Emerging technologies, the war in ukraine, and the risk of nuclear escalation. *Journal for Peace and Nuclear Disarmament, 6*(1), 28-46. https://doi.org/10.1080/25751654.2023.2219437

Fei, K., Zhou, J., Su, L., Wang, W., & Chen, Y. (2025). Log2Graph: A graph convolution neural network based method for insider threat detection. *Journal of Computer Security, 33* (1), 37-56.

Fox, D. B., McCollum, C. D., Arnoth, E. I., & Mak, D. J. (2018). Cyber wargaming: Framework for enhancing cyber wargaming with realistic business context. *MITRE CORP MCLEAN VAHOMELAND SECURITY SYSTEMS ENGINEERING AND DEVELOPMENT INSTITUTE.*

Frank, A. B. (2025). Gaming AI without AI. *The Journal of Defense Modeling and Simulation, 22* (1), 5-18.

Fu, N., Zhang, D., & Chen, Y. (2024). Detection and analysis of network system security based on machine learning. *Intelligent Decision Technologies*, (Preprint), 1-10.

Fujimoto, R. M. (2024). Development of the parallel and distributed simulation field. *SIMULATION, 100* (12), 1197-1223.

Galbraith, J. (2019). U.S. military undergoes restructuring to emphasize cyber and space capabilities. *The American Journal of International Law, 113* (3), 634–640. https://doi.org/10.1017/ajil.2019.39

Galbraith, J. (2020). United States creates the US space command and the US space force to strengthen military capabilities in space. *The American Journal of International Law, 114* (2), 323-326.

Giustozzi, F., Saunier, J., & Zanni-Merk, C. (2024). A semantic framework for condition monitoring in Industry 4.0 based on evolving knowledge bases. *Semantic Web, 15*(2), 583-611.

Goecks, V. G., Waytowich, N., Asher, D. E., Jun Park, S., Mittrick, M., Richardson, J., Vindiola, M., Logie, A., Dennison, M., Trout, T., Narayanan, P., & Kott, A. (2023). On games and simulators as a platform for development of artificial intelligence for command and control. *The Journal of Defense Modeling and Simulation: Applications, Methodology, Technology, 20*(4), 495-508. https://doi.org/10.1177/15485129221083278

Goldsmith, J. (Ed.). (2022). *The United States' Defend Forward Cyber Strategy: A Comprehensive Legal Assessment.* Oxford University Press.

Grimaila, M. R., Myers, J., Mills, R. F., & Peterson, G. (2012). Design and analysis of a dynamically configured log-based distributed security event detection methodology. *The Journal of Defense Modeling and Simulation, 9*(3), 219-241.

Gu, F., Xu, H., & He, D. (2024, September). How Does Variation in AI Performance Affect Trust in AI-infused Systems: A Case Study With In-Vehicle Voice Control Systems. In *Proceedings of the Human Factors and Ergonomics Society Annual Meeting* (Vol. 68, No. 1, pp. 1092-1097). Sage CA: Los Angeles, CA: SAGE Publications.

Guevara, C. (2025). Stock market trading via actor-critic reinforcement learning and adaptable data structure. *PeerJ. Computer Science, 11*, e2690. https://doi.org/10.7717/peerj-cs.2690

Gurmessa, D. K., & Jimma, W. (2025). A comprehensive evaluation of interpretable artificial intelligence for epileptic seizure diagnosis using an electroencephalogram: A systematic review. *Digital Health, 11*, 20552076251325411.

Hall, A., Murray, P., Boring, R. L., & Agarwal, V. (2024, September). Human-centered and explainable artificial intelligence in nuclear operations. In *Proceedings of the Human Factors and Ergonomics Society Annual Meeting* (Vol. 68, No. 1, pp. 1563-1568). Sage CA: Los Angeles, CA: SAGE Publications.

Hannay, J. E., van den Berg, T., Gallant, S., & Gupton, K. (2021). Modeling and Simulation as a Service infrastructure capabilities for discovery, composition, and execution of simulation services. *The Journal of Defense Modeling and Simulation, 18* (1), 5-28.

Hautz, J., Seidl, D., & Whittington, R. (2017). Open strategy: Dimensions, dilemmas, dynamics. *Long Range Planning, 50* (3), 298–309. https://doi.org/10.1016/j.lrp.2016.12.001

Hoey, J. (2025). Social organization as the collective management of uncertainty. *Collective Intelligence, 4*(1), 26339137251324131

Holburn, G. L. F., Maxwell, J. W., & Bonardi, J.-P. (2024). Safe Bets, Long Shots, and Toss-Ups: Strategic Engagements Between Activists and Firms. *Management Science, 70*(11), 7443–7462. https://doi.org/10.1287/mnsc.2022.01638

Huang, Q., Lynn, B. J., Dong, C., Ni, S., & Men, L. R. (2023). Relationship cultivation via social media during the COVID-19 pandemic: Evidence from China and the US. *International Journal of Business Communication, 60* (2), 512-542.

Hughes, C., & Turner, T. (2023). *Software Transparency: supply chain security in an era of a software-driven society.* John Wiley & Sons.

Ii, J. B. (2020). Russian interference in U.S. elections: How to protect critical election infrastructure from foreign participation. *Public Contract Law Journal, 49* (4), 709-734.

Iossa, E., Rey, P., & Waterson, M. (2022). Organising Competition for the Market. *Journal of the European Economic Association, 20*(2), 822-868. https://doi.org/10.1093/jeea/jvab044

Jain, N., Burman, E., & Mumovic, D. (2024). CIBSE TM54 energy projections III: A case study using dynamic simulation with detailed system modelling. *Building Services Engineering Research & Technology, 45*(4), 525-539.

Jayashree, K. Secureimagesec: A privacy-preserving framework for outsourced picture representation with content-based image retrieval. *Intelligent Data Analysis,* (Preprint), 1-22.

Jenefa, A., Vidhya, K., Taurshia, A., Naveen, V. E., Kuriakose, B. M., & Vijula, V. (2024). Enhancing distributed agent environments with quantum multi-agent systems and protocols. *Multiagent and Grid Systems, 20* (2), 109-127.

Jensen, B., Valeriano, B., & Whitt, S. (2024). How cyber operations can reduce escalation pressures: Evidence from an experimental wargame study. *Journal of Peace Research, 61* (1), 119-133. https://doi.org/10.1177/00223433231219440

Johnson, J. E., & Jordan, M. L. (2019). The development of undergraduate international business practicums in small business schools: An experiential learning framework. *Journal of Teaching in International Business, 30* (1), 6-32. https://doi.org/10.1080/08975930.2019.1628686

Jule, J. G. (2020). Workplace safety: a strategy for enterprise risk management. *Workplace health & safety, 68* (8), 360-365.

Jung, Y. J. (2024). Cyber shadows over nuclear peace: understanding and mitigating digital threats to global security. Journal of Asian Security and International Affairs, 11(2), 233-253.

Kakisim, A. G. (2024). A deep learning approach based on multi-view consensus for SQL injection detection. International Journal of Information Security, 23(2), 1541–1556. https://doi.org/10.1007/s10207-023-00791-y

Kanovich, M., Ban Kirigin, T., Nigam, V., Scedrov, A., & Talcott, C. (2017). Time, computational complexity, and probability in the analysis of distance-bounding protocols. *Journal of Computer Security, 25* (6), 585-630.

Karthika, K., & Rangasamy, D. P. (2024). Authorization of Aadhar data using Diffie Helman key with enhanced security concerns. *Journal of Intelligent & Fuzzy Systems, 46* (4), 8639-8658.

Katkuri, S. (2024). Need of Encryption Legislation: Protecting India's Digital Realm and Beyond. *Indian Journal of Public Administration, 70* (3), 562-578.

Katsarakes, E. A., Edwards, M., & Still, J. D. (2024, September). Where Do Users Look When Deciding If a Text Message is Safe or Malicious? In *Proceedings of the Human Factors and Ergonomics Society Annual Meeting* (Vol. 68, No. 1, pp. 221-225). Sage CA: Los Angeles, CA: SAGE Publications.

Khan, A. M., & Khan, A. A. (2025). Cyber-deterrence and Cyber-CBMs: Way Forward for Managing India-Pakistan Cyber

Rivalry. *Journal of Asian and African Studies*, 00219096251332932.

Kommidi, V. R., Padakanti, S., & Pendyala, V. (2024). Securing the Cloud: A Comprehensive Analysis of Data Protection and Regulatory Compliance in Rule-Based Eligibility Systems. *Technology (IJRCAIT)*, 7(2).

Kurteva, A., Chhetri, T. R., Pandit, H. J., & Fensel, A. (2024). Consent through the lens of semantics: State of the art survey and best practices. *Semantic Web*, *15*(3), 647-673.

Lawrence, C. (2025). Gathering around a satellite image: Visual media cycles of the nuclear nonproliferation complex. *Social Studies of Science*, *55*(1), 3-36.

Lazarova-Molnar, S., Friederich, J., & Niloofar, P. (2024). Reliability modeling and simulation: advancements with data-driven techniques and expert knowledge integration. *SIMULATION*, 00375497241300372.

Lee, J. (2025). Effects of Fair Value Reporting of Derivatives on Liquidity Management Policies and Firm Value: Evidence From SFAS No. 133. *Journal of Accounting, Auditing & Finance*, *40*(1), 76-100.

Lee, M. D. (2024). Using cognitive models to improve the wisdom of the crowd. *Current Directions in Psychological Science*, *33*(5), 308-316.

Lee, O., Currano, R., Miller, D., Yim, S., & Sirkin, D. (2024, September). Improving Situation Awareness in Partially Automated Vehicles: The Effectiveness of Redesigned Visual Signals. In *Proceedings of the Human Factors and*

Ergonomics Society Annual Meeting (Vol. 68, No. 1, pp. 888-893). Sage CA: Los Angeles, CA: SAGE Publications.

Leeuw, F. L., & Leeuw, B. (2012). Cyber society and digital policies: Challenges to evaluation? *Evaluation, 18* (1), 111-127.

Leong, M., Abdelhalim, A., Ha, J., Patterson, D., Pincus, G. L., Harris, A. B., Eichler, M., & Zhao, J. (2024). MetRoBERTa: Leveraging Traditional Customer Relationship Management Data to Develop a Transit-Topic-Aware Language Model. *Transportation Research Record: Journal of the Transportation Research Board, 2678* (9), 215-229. https://doi.org/10.1177/03611981231225655

Li, C. (2019). A repeated call for omnibus federal cybersecurity law. *The Notre Dame Law Review, 94* (5), 2211.

Li, L., Xiao, B., & Wu, X. (2025). Optimal control of strength allocation strategies generation with complex constraints. *Transactions of the Institute of Measurement and Control, 47* (4), 634-646.

Li, Y. (2018). Including a Definition of Operation of Law in the Federal Acquisition Regulation: A Roadmap for Government Contractors Engaging in Merger and Acquisition Transactions. *Pub. Cont. LJ, 48*, 819.

Li, X., Khan, L., Zamani, M., Wickramasuriya, S., Hamlen, K., & Thuraisingham, B. (2023). Con2Mix: A semi-supervised method for imbalanced tabular security data. *Journal of Computer Security, 31* (6), 705-726.

Li, X., Wang, J., Sun, L., & Li, W. (2024). A self-supervised feature fusion approach to situation assessment. *Journal of Intelligent & Fuzzy Systems, 4 7*(5-6), 487-497.

Lin-Greenberg, E., Pauly, R. B., & Schneider, J. G. (2022). Wargaming for international relations research. *European Journal of International Relations, 28* (1), 83-109.

Liu, H., Wang, Y., & Wang, K. (2024). Adaptive secure lateral control of autonomous electric vehicles under denial-of-service attacks. *Transactions of the Institute of Measurement and Control, 46* (13), 2558-2569.

Liu, L., Zhou, Y., Xu, Q., Shi, Q., & Hu, X. (2023). Improved technique for order of preference by similarity to ideal solution method for identifying key terrain in cyberspace asset layer. *PloS One, 18* (7), e0288293-e0288293.https://doi.org/10.1371/journal.pone.0288293

Liu, X., Wu, Y., Fiumara, G., & De Meo, P. (2024). Heterogeneous graph community detection method based on K-nearest neighbor graph neural network. *Intelligent Data Analysis, 28* (6), 1445-1466.

Lo, P. (2024, September). Healthcare innovations: Enhancing patient privacy and security in the digital era. In *Healthcare Management Forum* (Vol. 37, No. 5, pp. 363-365). Sage CA: Los Angeles, CA: SAGE Publications.

Luo, R., Zhao, S., Kuck, J., Ivanovic, B., Savarese, S., Schmerling, E., & Pavone, M. (2024). Sample-efficient safety assurances using conformal prediction. *The International Journal of Robotics Research, 43* (9), 1409-1424.

Lushenko, P., & Sparrow, R. (2024). Artificial intelligence and US military cadets' attitudes about future war. *Armed Forces & Society,* 0095327X241284264.

Madhira, N., Pelletier, J. M., Johnson, D., & Mishra, S. (2024). Code red: A nuclear nightmare-navigating ransomware response at an eastern european power plant. *Journal of* Information *Technology Teaching Cases, 14* (1), 108–118. https://doi.org/10.1177/20438869231155934

Maggu, P., Singh, S., Sinha, A., Biamba, C. N., Iwendi, C., & Hashmi, A. (2025). Sustainable and optimized power solution using hybrid energy system. *Energy Exploration & Exploitation, 43* (2), 526-563.

Mahbod, R., Fallon, G., & Hinton, D. (2019). Cybersecurity challenges and application within the army national guard. *The Armed Forces Comptroller, 64* (4), 38-45.

Mensah, N. K., Adzakpah, G., Kissi, J., Taylor-Abdulai, H., Johnson, S. B., Agbeshie, P. A.,

Opoku, C., Abakah, J., Osei, E., Agyekum, A. Y., & Boadu, R. O. (2024). Health Professionals' Ethical, Security, and Patient Safety Concerns Using Digital Health Technologies: A Mixed Method Research Study. *Health Services Insights, 17.* https://doi.org/10.1177/11786329241303379

Mensen, J. M., Holland, S. B., Helton, W. S., Shaw, T. H., & Peterson, M. S. (2024). Prolonging the response movement reduces commission errors in a high-go, low-no-go target detection task and composite metrics of performance miss this effect. *Human factors, 66* (4), 1118-1131.

Mentzas, G., Fikardos, M., Lepenioti, K., & Apostolou, D. (2024). Exploring the landscape of trustworthy artificial intelligence: Status and challenges. *Intelligent Decision Technologies, 18* (2), 837-854.

Mergel, I. (2024). Social affordances of agile governance. *Public Administration Review, 84* (5), 932–947. https://doi.org/10.1111/puar.13787

Moghadasi, M., Shirmohammadi, M., & Ghasemi, A. (2024). A framework for evaluation of crowdsourcing platforms performance. *Information Development, 40*(4), 635-647.

Moradi, M., & Li, Q. (2021). Rogue people: On adversarial crowdsourcing in the context of cyber security. *Journal of Information, Communication & Ethics in Society (Online), 19* (1), 87-103. https://doi.org/10.1108/JICES-08-2019-0100

Moradi, M., Weng, Y., & Lai, Y. (2022). Defending smart electrical power grids against cyberattacks with deep Q -learning. *PRX Energy, 1*(3), 033005. https://doi.org/10.1103/PRXEnergy.1.033005

Moses, M., Pearl, G. A., & Caio, S. (2025). Product innovation and SME export intensity: The moderating role of government support business environment and quality certification. *The International Journal of Entrepreneurship and Innovation,* 14657503251332139.

Muthukumar, S., & Ashfauk Ahamed, A. K. (2024). A novel framework of DDoS attack detection in network using hybrid heuristic deep learning approaches with attention mechanism. *Journal of High Speed Networks, 30*(2), 251-277.

Nayak, B., Bhattacharyya, S. S., & Krishnamoorthy, B. (2022). Customer value creation—a case study of Indian health insurance industry from value net perspective. *Journal of Health Management, 24* (4), 539-555.

Neculcea, C. (2023). China's Three Warfare Strategy. Origins, Evolution, Applicability. *Journal of Defense Resources Management (JoDRM)*, 14(1), 19–30.

Nieuwenhuizen, E. (2024). Algorithm Registers: A Box-Ticking Exercise or Meaningful Tool for Transparency?. *Information Polity, 29* (4), 415-433.

Oh, S. H., Jeong, M. K., Kim, H. C., & Park, J. (2023). Applying reinforcement learning for enhanced cybersecurity against adversarial simulation. *Sensors (Basel, Switzerland), 23*(6), 3000. https://doi.org/10.3390/s23063000

Oppenheimer, H. (2024). How the process of discovering cyberattacks biases our understanding of cybersecurity. *Journal of Peace Research, 61* (1), 28-43.

Pagale, M., Sharma, R., & Thakare, A. (2025). A review for autonomous vehicles technologies. *Multiagent and Grid Systems*, 15741702241296428.

Papworth, J., & Dence, R. (2024). From warships to whaleships: Former Royal Navy vessels entering the South Seas fishery in the post-Napoleonic period, 1815–1845. *International Journal of Maritime History, 36* (4), 842-873.

Payton, T., & Claypoole, T. (2023). *Privacy in the age of Big data: Recognizing threats, defending your rights, and protecting your family.* Rowman & Littlefield.

Perdana, A., Aminanto, M. E., & Anggorojati, B. (2024). Hack, heist, and havoc: The Lazarus Group's triple threat to global cybersecurity. *Journal of Information Technology Teaching Cases*, 20438869241303941.

Perla, P. (2022). Wargaming and the cycle of research and learning. *Scandinavian Journal of Military Studies, 5*(1), 197–208. https://doi.org/10.31374/sjms.124

Perron, P. (2023). Moving beyond the sanctuary paradigm: Canada must face up to the reality of a contested and dangerous space environment. *International Journal, 78*(1-2), 147-171.

Phillips, W., Roehrich, J. K., Kapletia, D., & Alexander, E. (2022). Global value chain reconfiguration and COVID-19: Investigating the case for more resilient redistributed models of production. *California Management Review, 64*(2), 71-96.

Plotnikov, M., & Collura, J. (2022). Integrating unmanned aircraft systems into state department of transportation highway bridge inspection procedures: challenges, implications, and lessons learned. *Transportation research record, 2676*(2), 529-540

Pournelle, P. (2024). The need for cooperation between wargaming and modeling & simulation for examining cyber, space, electronic warfare, and other topics. *Journal of Defense Modeling and Simulation, 21*(4), 359–362. https://doi.org/10.1177/15485129221118100

Preethichandra, D. M. G., Piyathilaka, L., Sul, J., Izhar, U., Samarasinghe, R., Arachchige, S. D., & de Silva, L. C. (2024). Passive and active exoskeleton solutions: Sensors, actuators, applications, and recent trends. *Sensors (Basel, Switzerland).*

Qi, G. (2024). Theoretical reconstruction and practical dimension of labor education in digital age. *Journal of Computational Methods in Science and Engineering,* 14727978241299626.

Qureshi, A., Marvi, M., Shamsi, J. A., & Aijaz, A. (2022). eUF: A framework for detecting over-the-air malicious updates in autonomous vehicles. *Journal of King Saud University. Computer and Information Sciences, 34* (8), 5456–5467. https://doi.org/10.1016/j.jksuci.2021.05.005

Radanliev, P., Santos, O., & Brandon-Jones, A. (2024). Capability hardware enhanced instructions and artificial intelligence bill of materials in trustworthy artificial intelligence systems: analyzing cybersecurity threats, exploits, and vulnerabilities in new software bills of materials with artificial intelligence. *The Journal of Defense Modeling and Simulation,* 15485129241267919.

Radmanesh, M., Kumar, M., Nemati, A., & Sarim, M. (2016). Dynamic optimal UAV trajectory planning in the national airspace system via mixed integer linear programming. *Proceedings of the Institution of Mechanical Engineers, Part G: Journal of Aerospace Engineering, 230*(9), 1668-1682.

Parkin, S., & Chua, Y. T. (2022). A cyber-risk framework for coordination of the prevention and preservation of behaviours. *Journal of Computer Security, 30* (3), 327-356.

Rathore, R. S., Hewage, C., Kaiwartya, O., & Lloret, J. (2022). In-vehicle communication cyber security: Challenges and solutions. *Sensors (Basel, Switzerland), 22* (17), 6679. https://doi.org/10.3390/s22176679

Reddie, A. W., & Goldblum, B. L. (2023). Evidence of the unthinkable: Experimental wargaming at the nuclear threshold. *Journal of Peace Research, 60* (5), 760–776. https://doi.org/10.1177/00223433221094734

Riggs, H., Tufail, S., Parvez, I., Tariq, M., Khan, M. A., Amir, A., Vuda, K. V., & Sarwat, A. I. (2023). Impact, vulnerabilities, and mitigation strategies for cyber-secure critical infrastructure. *Sensors (Basel, Switzerland), 23*(8), 4060. https://doi.org/10.3390/s23084060

Reid, F., Pravinkumar, S. J., Maguire, R., Main, A., McCartney, H., Winters, L., & Dong, F. (2025). Using machine learning to identify frequent attendance at accident and emergency services in Lanarkshire. *Digital Health, 11,* 20552076251315293

Petty, M. D., Bland, J. A., Whitaker, T. S., Cantrell, W. A., Maxwell, K. P., Colvett, C. D., & Bearss, E. M. (2024). Simulating cyberattacks with extended Petri nets. *SIMULATION, 100*(12), 1257-1280.

Reuer, J. J. (2024). Revisiting research on the governance of interorganizational relationships. *Journal of Management Scientific Reports, 2*(3-4), 267-279.

Prasath, J. S., Shyja, V. I., Chandrakanth, P., Kumar, B. K., & Raja Basha, A. (2024). An optimal secure defense mechanism for DDoS attack in IoT network using feature optimization and intrusion detection system. *Journal of Intelligent & Fuzzy Systems, 46*(3), 6517-6534.

Rühlig, T. N., & Ten Brink, T. (2021). The Externalization of China's Technical Standardization Approach. *Development & Change, 52*(5), 1196–1221. https://doi.org/10.1111/dech.12685

Ryseff, J., & Bond, M. (2025). Small is beautiful. *The Journal of Defense Modeling and Simulation, 22*(1), 19-23.

Sabidi, M. L., & Zolkipli, M. (2024). The Role of Risk Management in Cybersecurity Protocols. *Borneo International Journal,* 7(2), 77–81. https://majmuah.com/journal/index.php/bij/article/view/643

Sanders, G., Jang, W. J., & Holderness, A. (2022). *Defense acquisition trends 2021.* Rowman & Littlefield.

Sarjakivi, P., Ihanus, J., & Moilanen, P. (2024). Using Wargaming to Model Cyber Defense Decision-Making: Observation-Based Research in Locked Shields. In *Proceedings of the European Conference on Cyber Warfare and Security* (Vol. 23, No. 1). Academic Conferences International Ltd.

Sarkar, S., Kakade, K., & AK, S. (2024). Utilising blockchain technology to implement a security control method for node access to the Internet of Things. *Intelligent Decision Technologies, 18* (2), 953-963.

Sattar, S., Hulsey, A., Hagen, G., Naeim, F., & McCabe, S. (2021). Implementing the performance-based seismic design for new reinforced concrete structures: Comparison among ASCE/SEI 41, TBI, and LATBSDC. *Earthquake Spectra, 37* (3), 2150-2173

Schechter, B., Schneider, J., & Shaffer, R. (2021). Wargaming as a methodology: The international crisis wargame and experimental wargaming. *Simulation & Gaming, 52*(4), 513–526. https://doi.org/10.1177/1046878120987581

Schwarz, J. O. (2020). Revisiting scenario planning and business wargaming from an open strategy perspective. *World Futures Review, 12* (3), 291-303. https://doi.org/10.1177/1946756720953182

Schwarz, J. O., Ram, C., & Rohrbeck, R. (2019). Combining scenario planning and business wargaming to better anticipate future competitive dynamics. *Futures: The Journal of Policy, Planning and Futures Studies*, 105, 133–142. https://doi.org/10.1016/j.futures.2018.10.001

Segal, A. (2020). China's pursuit of cyberpower. *Asia Policy, 15* (2), 60-66.

Segate, R. V. (2024). Drafting a Cybersecurity Standard for Outer Space Missions: On Critical Infrastructure, China, and the Indispensability of a Global Inclusive Approach. *Journal of Asian Security and International Affairs, 11*(3), 345-375.

Selvarajan, S., Shankar, A., Uddin, M., Alqahtani, A. S., Al, S. T., & Viriyasitavat, W. (2025). A smart decentralized identifiable distributed ledger technology-based blockchainn (DIDLT-BC) model for cloud-IoT security. Expert Systems, 42(1), 1–25. https://doi.org/10.1111/exsy.13544

Sethuraman, A. V., Sheppard, A., Bagoren, O., Pinnow, C., Anderson, J., Havens, T. C., & Skinner, K. A. (2024). Machine learning for shipwreck segmentation from side scan sonar imagery: Dataset and benchmark. *The International Journal of Robotics Research*, 02783649241266853.

Settembre-Blundo, D., González-Sánchez, R., Medina-Salgado, S., & García-Muiña, F. E. (2021). Flexibility and resilience in corporate decision making: a new sustainability-based risk management system in uncertain times. *Global Journal of Flexible Systems Management, 22* (Suppl 2), 107-132.

Shao, C., & Li, Y. (2022). Multistage Attack–Defense Graph Game Analysis for Protection Resources Allocation Optimization

Against Cyber Attacks Considering Rationality Evolution. Risk Analysis: An International Journal, 42(5), 1086–1105. https://doi.org/10.1111/risa.13837

Sharma, M. (2024). Risks of Cyber Security Threats, Cyber Terrorism and Cyber Warfare: An Analysis of Impact and Countermeasures. *Risks of Cyber Security Threats, Cyber Terrorism and Cyber Warfare: An Analysis of Impact and Countermeasures* (July 31, 2024).

Sharmin, S., Mansor, H., Abdul Kadir, A. F., & Aziz, N. A. (2024). Benchmarking frameworks and comparative studies of Controller Area Network (CAN) intrusion detection systems: A review. *Journal of Computer Security, 32* (5), 477-507.

Sheikh, Z. A., Singh, Y., Singh, P. K., & Gonçalves, P. J. S. (2023). Defending the defender: Adversarial learning based defending strategy for learning based security methods in cyber-physical systems (CPS). *Sensors (Basel, Switzerland), 23* (12), 5459. https://doi.org/10.3390/s2312545

Shukla, S., & Patel, S. J. (2022). A novel ECC-based provably secure and privacy-preserving multi-factor authentication protocol for cloud computing. *Computing, 104* (5), 1173–1202. https://doi.org/10.1007/s00607-021-01041-6

Sisson, M. (2021). Combining Wargaming and Simulation Analysis. In *Simulation and Wargaming* (pp. 183–202).

Skowronek, E. O., Baker, L. S., Ahmed, R., & Marvi, H. (2025). Model-predictive optimal control of ferrofluidic microrobots in three-dimensional space. *The International Journal of Robotics Research, 44* (5), 826-839.

Skraaning Jr, G., & Jamieson, G. A. (2024). The failure to grasp automation failure. *Journal of Cognitive Engineering and Decision Making, 18*(4), 274-285.

Slayton, R. (2021). Governing uncertainty or uncertain governance? Information security and the challenge of cutting ties. *Science, Technology, & Human Values, 46* (1), 81-111.

Smeets, M. (2018). The strategic promise of offensive cyber operations. *Strategic Studies Quarterly: SSQ, 12* (3), 90-113.

Smith, C. (2024). Chinese Lawfare in Conflict: The Threat to US Operations (FORTHCOMING). *Available at SSRN 5052517.*

Sodhi, I. S., Kaur, M., & Kaur, J. (2024). Policy and Infrastructure for Data Privacy and Data Protection in South Indian States: Issues and Prospects. *Indian Journal of Public Administration, 70* (3), 593-608.

Sohrabi, S., Katz, M., Hassanzadeh, O., Udrea, O., Feblowitz, M. D., & Riabov, A. (2019). IBM scenario planning advisor: Plan recognition as AI planning in practice. *Ai Communications, 32* (1), 1-13.

Sohrabinejad, A., Hafshejani, K. F., & Sobhani, F. M. (2025). Leveraging genetic algorithms and system dynamics for effective multi-objective policy optimization: a case study on the broadcasting industry. *SIMULATION, 101*(4), 453-476.

Spaniol, M. J. (2024). Organizing foresight tools. *World Futures Review, 16* (3), 261-276

Stebbins, D. M. (2023). *Developing a Narrative Assessment Framework to Enable Learning Within US Department of*

Defense Wargaming (Doctoral dissertation, George Mason University).

Stensrud, R., & Valaker, S. (2023, September). Design of a trusted shift of Coordination Forms: Supporting Collaboration to handle future non-human intelligent collaborators (NICs). In *Proceedings of the Human Factors and Ergonomics Society Annual Meeting* (Vol. 67, No. 1, pp. 1651-1659). Sage CA: Los Angeles, CA: SAGE Publications

Storbeck, M., Jacobs, G., Schuilenburg, M., & van den Akker, R. (2025). Surveillance experiences of extinction rebellion activists and police: Unpacking the technologization of Dutch protest policing. *Big Data & Society, 12* (1), 20539517241307892.

Sullivan, G. (2020). The Kaspersky, ZTE, and Huawei sagas: Why the United States is in desperate need of a standardized method for banning foreign federal contractors. *Public Contract Law Journal,* 49(2), 323-349.

Swallow, R. C. (2023). Considering the cost of cyber warfare: advancing cyber warfare analytics to better assess tradeoffs in system destruction warfare. *The Journal of Defense Modeling and Simulation, 20* (1), 3-37.

Tan, Y. (2024). Implications of blockchain-powered marketplace of preowned virtual goods. *Production and Operations Management, 33* (6), 1393-1409.

Tarraf, D. C., Gilmore, J. M., Barnett, D. S., Boston, S., Frelinger, D. R., Gonzales, D., Hou, A. C., & Whitehead, P. (2025). An experiment in tactical wargaming with platforms enabled by artificial intelligence. *The Journal of Defense Modeling and*

Simulation: Applications, Methodology, Technology, 22 (1), 59-76. https://doi.org/10.1177/15485129221097103

TM, T., & Taurshia, A. (n.d.). DeepGAN: Utilizing generative adversarial networks for improved deep learning. *International Journal of Knowledge-based and Intelligent Engineering Systems*, (Preprint), 1-17.

Toga, A. W., Neu, S., Sheehan, S. T., Crawford, K., & Alzheimer's Disease Neuroimaging Initiative. (2024). The informatics of ADNI. *Alzheimer's & Dementia, 20* (10), 7320-7330.

Topor, L. (2024). Cyber Warfare: Global Trends and Proxy Wars. In *Cyber Sovereignty: International Security, Mass Communication, and the Future of the Internet* (pp. 75–110). Cham: Springer Nature Switzerland.

Tran, H. T., Domerçant, J. C., & Mavris, D. N. (2015). Evaluating the agility of adaptive command and control networks from a cyber complex adaptive systems perspective. *The Journal of Defense Modeling and Simulation, 12* (4), 405-422.

Tribou, K. J., Roberson, M., & Roberson, M. (2025). A Risky Proposition: Setting Objectives and Assessing Business Environment Risks Using the Enterprise Risk Management Framework. *Issues in Accounting Education, 40* (1), 173–201. https://doi.org/10.2308/ISSUES-2023-043

Trigg, H. B., & Mrozowski, S. A. (2024). Dominion and improvement: The moral ecologies of colonial encounters. *Journal of Social Archaeology, 24* (3), 246-265.

Trim, P. R. J., & Lee, Y. (2021). The global cyber security model: Counteracting cyber-attacks through a resilient partnership

arrangement. *Big Data and Cognitive Computing, 5* (3), 32
https://doi.org/10.3390/bdcc5030032

Trim, P., & Lee, Y. I. (2025). *Cyber Security Management and
Strategic Intelligence.* Taylor & Francis.

Trope, R. L. (2019). To secure, or not secure, data integrity - that is
the question: Cybersecurity developments. *The Business
Lawyer, 75* (1), 1655.

Trope, R. L. (2020). To secure, or not secure, data integrity-that is
the question: Cybersecurity developments. *The Business
Lawyer, 75* (1), 1655-1666.

Tyagi, G., Behera, L. K., Yadav, A., & Verma, N. K. (2024).
Guarding Against China's Predatory Acquisition of Foreign
Technologies: A Quad Perspective. *Science, Technology and
Society, 29* (3), 435-453.

Uysal, S., & Sandıkkaya, M. T. (2025). A survey on obstacles to the
widespread use of connected and automated vehicles. *Journal
of Ambient Intelligence and Smart Environments, 17* (1), 28-
43.

Vedral, B. (2021, May). The vulnerability of the financial system to a
systemic cyberattack. In *2021 13th International Conference
on Cyber Conflict (CyCon)* (pp. 95–110). IEEE.

Van Rooyen, P., Tapinos, E., & Revie, M. (2025). Strategic Foresight
in Fintech: Harnessing Scenario Planning for Future
Readiness.

Wang, A. Z., Borland, D., & Gotz, D. (2025). A framework to
improve causal inferences from visualizations using
counterfactual operators. *Information Visualization, 24*(1), 24-
41.

Wang, H., Lau, N., & Gerdes, R. M. (2018). Examining cybersecurity of cyberphysical systems for critical infrastructures through work domain analysis. *Human factors, 60* (5), 699-718.

Wang, S., & Khan, A. (2025). Exploring the factors driving the sustainable consumer intentions for over-the-air updates in electric vehicles. *Energy Exploration & Exploitation, 43* (1), 319-339.

Wei, F., Xu, N., Huang, S., & Cao, Y. (2025). Disturbance observer–based adaptive neural finite-time control for nonstrict-feedback nonlinear systems with input delay. *Transactions of the Institute of Measurement and Control, 47* (6), 1172-1187.

West, J., Chu, M., Crooks, L., & Bradley-Ho, M. (2018). Strategy war games: How business can outperform the competition. *The Journal of Business Strategy, 39* (6), 3-12. https://doi.org/10.1108/JBS-11-2017-0154

Winger, G. H. (2023). *Cybersecurity in the U.S.-Philippine alliance: Mission seep. The Pacific Review, 36*(6), 1365–1393.

Wong, Y. H., Ryseff, J., & Riggs, N. (2025). Artificial intelligence and wargaming. *The Journal of Defense Modeling and Simulation, 22*(1), 3-4.

Woods, D. W., Böhme, R., Wolff, J., & Schwarcz, D. (2023). Lessons lost: Incident response in the age of cyber insurance and breach attorneys. In *The 32nd USENIX Security Symposium (USENIX Security 23)* (pp. 2259–2273).

Xu, J. (2024). Refining and implementing of a decision tree based risk assessment model for college students' innovation and entrepreneurship. *Journal of Computational Methods in Science and Engineering, 24* (4-5), 3093-3111.

Xu, Y., Huang, Q., Teng, S., Wang, Y., Sun, H., Li, M., Li, J., Zhu, M., & Tang, X. (2025). Aged smart-care application program for promoting quality of life among older adults in the community: Study protocol of a three-arm randomized controlled trial. *DIGITAL HEALTH, 11*. https://doi.org/10.1177/20552076251326218

Xu, Z., Hu, Z., Zheng, X., Zhang, H., & Luo, Y. (2023). A matrix factorization recommendation model for tourism points of interest based on interest shift and differential privacy. *Journal of Intelligent & Fuzzy Systems, 44* (1), 713-727.

Ye, M., Yan, X., Chen, N., & Liu, Y. (2024). A robust multi-scale learning network with quasi-hyperbolic momentum-based Adam optimizer for bearing intelligent fault diagnosis under sample imbalance scenarios and strong noise environment. *Structural Health Monitoring, 23* (3), 1664-1686.

Yoon, K., Kogan, A., Vasarhelyi, M. A., & Pearce, T. (2024). External NonfinancialMeasures in Substantive Analytical Procedures: Contributions of Weather Information. *Journal of Information Systems, 38*(2), 143–162. https://doi.org/10.2308/ISYS-2023-066

Young, S., Gentile, G. P., Sacks, B. J., Edenfield, N., & Givens, A. (2023). *A History of the Strategic Implications of the Great Recession and Its Aftermath on US National Defense* (p. 214). RAND Corporation.

Yu, Y., Li, Q., Duan, M., Yuan, M., & Song, Z. (2024). Traffic and transportation management data storage terminal based on Internet of Things. *Intelligent Decision Technologies*, (Preprint), 1-15.

Yusuf, S. A., Khan, A., & Souissi, R. (2024). Vehicle-to-everything (V2X) in the autonomous vehicles domain – A technical review of communication, sensor, and AI technologies for road user safety. *Transportation Research Interdisciplinary Perspectives, 23*, 100980. https://doi.org/10.1016/j.trip.2023.100980

Zhang, Z., Guo, Q., Grigorev, M. A., & Kholodilin, I. (2024). Construction method of a digital-twin simulation system for SCARA robots based on modular communication. *Sensors (Basel, Switzerland), 24* (22), 7183. https://doi.org/10.3390/s24227183

Zhen, S., Ma, M., Liu, X., Chen, F., Zhao, H., & Chen, Y. H. (2023). Model-based robust control design and experimental validation of SCARA robot system with uncertainty. *Journal of Vibration and Control, 29* (1-2), 91-104.

Zhukabayeva, T., Zholshiyeva, L., Karabayev, N., Khan, S., & Alnazzawi, N. (2025). Cybersecurity solutions for industrial internet of things-edge computing integration: Challenges, threats, and future directions. *Sensors (Basel, Switzerland), 25* (1), 213. https://doi.org/10.3390/s25010213

www.ingramcontent.com/pod-product-compliance
Lightning Source LLC
Chambersburg PA
CBHW071228210326
41597CB00016B/1989